Power System State Estimation

For a complete listing of titles in the
Artech House Power Engineering Series,
turn to the back of this book.

Power System State Estimation

Mukhtar Ahmad

ARTECH
HOUSE

BOSTON | LONDON
artechhouse.com

Library of Congress Cataloging-in-Publication Data
A catalog record for this book is available from the U.S. Library of Congress.

British Library Cataloguing in Publication Data
A catalogue record for this book is available from the British Library.

Cover design by Vicki Kane

ISBN 13: 978-1-60807-511-9

685 Canton Street
Norwood, MA 02062

10 9 8 7 6 5 4 3 2 1

Contents

Preface

This book is written primarily to introduce to reader the developments in the field of state estimation of power system that have taken place during last 40 years. It is intended for users with background in power system, some basic control theory, and linear algebra. Familiarity with basic statistics is helpful in understanding the book. This book will be useful to power engineers working in control centers of electric utilities. I have also considered the requirement of graduate students of power engineering who want to go for research in the area of state estimation in power system.

State estimation is one of the most important functions in a power system operation and control. Schweppe and others in 1970s proposed the application of state vector, voltage magnitude, and phase angle in the control of power system. The knowledge of state vector enables the control centers to take important decisions about security and contingency evaluation.

In a power system the information available from various measuring equipments located at different places may not always be available, and the measurement errors in these instruments cannot be ruled out. Also the network impedances and topology is not accurately known. The job of state estimation is to estimate the state variables from set of redundant measurements. With the developments in smart grid, availability of phasor measurement units, and integration of renewable energy sources, the requirements of the state estimation system are changing. The modern distribution systems are also making use of state estimation in managing the power distribution systems efficiently. Thus, a need was felt by the author to write a book that can include these new developments, apart from the conventional theory of state estimation.

This book covers various aspects of state estimation in nine chapters. Chapter 1 describes the role of state estimation in energy management system. This chapter introduces the reader to the importance of state estimation in power system security. Chapter 2 gives the basic idea of power flow analysis in power system This chapter has been included to familiarize the reader with importance of states (voltage magnitude and angle) in a power system. The basic difference between power flow analysis and state estimation is also discussed. Chapter 3 deals with the state estimation theory and most commonly used methods of state estimation.

In state estimation topology of the network plays an important part. Chapter 4 is devoted to network observability based on the topology of the network. Chapters 5 and 6 introduce the topics of bad data detection and elimination. In Chapter 6 state estimation programs that work well even in the presence of bad data (called robust estimators) are discussed.

Chapter 7 is added to the book so that the application of current measurement in state estimation can be understood. Chapter 8 describes an important development in state estimation with the availability of phasor measurement units (PMUs). The application of PMUs, their optimal location, and their application in bad data processing is described. Presently, there is growing interest in application of PMUs to monitor the state of power system. Thus, this chapter is likely to be used by researchers and power engineers working in this area. Finally, Chapter 9 is about state estimation in distribution system. Again this is an area in which not much literature is available.

The author wishes to express his sincere gratitude to Mrs. Salma Shaheen, associate professor in the Women's Polytechnic AMU, for her help thoroughly reading the manuscript and giving suggestions for improvement. I am also thankful to the reviewer of this book who made constructive suggestions in improving almost all the chapters.

1

Energy Management Systems

In any developed country, more than one-third of energy consumed is in the form of electrical energy. The modern-day power systems are responsible for generating, transmitting, and distributing electric energy. To manage such large and complex systems, computer-based monitoring and control is essential. The management system for computer control of a power system is known as a *control center* or *energy management system.*

A modern energy management system performs many functions, such as automatic generation control, system security functions, and economic scheduling [1–3]. The functional diagram of a modern energy system is shown in Figure 1.1. As can be seen from this diagram, the functions of energy management system can be grouped into three categories:

1. Data acquisition and processing;
2. Energy/economy functions;
3. Security monitoring and control.

In this book we are mainly concerned with the role of state estimation in power system security monitoring and control.

1.1 Real-Time Control of a Power System

Modern day power systems are divided into various areas. Each of these areas is generally interconnected to its neighboring areas. The electric power systems interconnect because the interconnected systems are more reliable.

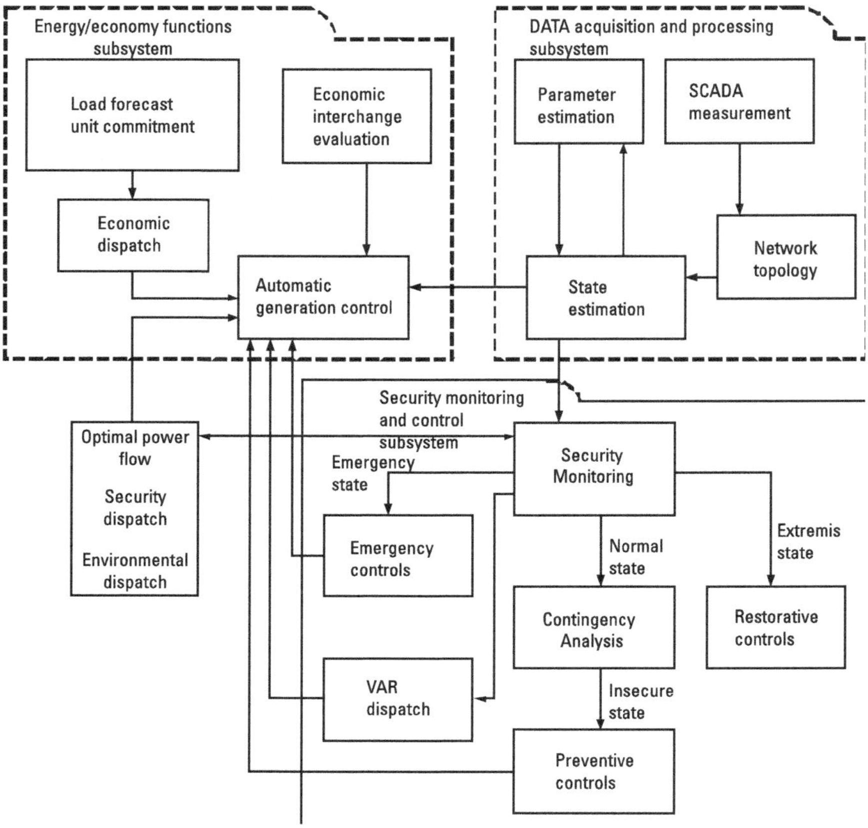

Figure 1.1 Functional diagram of a modern energy management system.

The transmission lines that connect an area to its neighboring area are called tie lines. Power sharing between two areas occurs through these tie lines. In interconnected systems it is important to keep the frequency the same in all the areas. The frequency can be kept constant under steady state conditions only. However, power systems are continuously subject to disturbances covering a wide range of conditions. Typical examples of disturbance conditions are sudden changes in load demand, customer load changes, losses of one or more transmission lines, modifications in the system configuration, equipment outages, and generator failures. In modern power systems, renewable energy sources of power generation such as wind power and solar energy are connected to conventional systems. Because of the variable nature of power generation by these sources, and if they are present in significant proportions, they also contribute to disturbances in the power system [4]. Power system security may be explained as the probability of the system operating point remaining within given acceptable ranges under conditions of contingencies. It is a function of time and of the robustness of the system with respect to imminent disturbances. The notion of security is the basis of all real-time monitoring and control in today's power systems. The working definition of security is in terms of the system state. The system state is a compact description used to summarize key information about the system; once the system state is known, it can be used to express any variables of interest.

A state in terms of power system security is classified as being one of three or four possible conditions: emergency, restorative, or normal. A *normal* state is classified when all operating constraints and load constraints are satisfied. An *emergency* state is one in which one or more of the physical operating limits are violated (e.g., line overloads, over/under voltages, over/under frequency). A *restorative* state is one where one or more of the loads are not met (partial or total blackout), but the *partial* system is operating in a normal state. The fourth state of the system is a condition when the power system loses stability and results in total blackout. The notion of security is defined with respect to a set of *credible* contingencies. A normal state is *secure* if all postulated contingencies result in secure *normal* operations; if a disturbance transitions the state into emergency, the state is *insecure* with respect to that contingency. This definition is well aligned with the intuitive notion that a secure power system is one that has low probability of blackout or equipment damage. The principal role of power system control is to maintain a secure system state (i.e., to prevent the system state from transitioning from secure to emergency over the widest range of operating conditions).

As the demands for reliable electric power became greater, and as labor became a more significant part of the cost of providing electric power, technologies known as supervisory control and data acquisition (SCADA) systems were developed that would allow remote monitoring and even control of key system parameters. SCADA systems began to reduce and even eliminate the need for personnel to be on hand at substations.

Early SCADA systems provided remote indication and control of substation parameters using technology borrowed from automatic telephone switching systems. Data rates on these early systems were slow—data was sent in the same manner as rotary-dial telephone commands, at 10 bits per second—so only a limited amount of information could be passed using this technology. Presently a full-fledged SCADA system is made up of signal hardware for input/output, networks, control equipment, user interface (sometimes called the human-machine interface or HMI), communication equipment, and the software to go with it all. The central system of SCADA is often miles away from where the operations take place. Thus, the system also needs on-site sensors to collect and monitor data.

The measurement and control system of SCADA has one master terminal unit (MTU), which could be called the brain of the system, and a large number of remote terminal units (RTUs). The RTUs gather the data locally and send them to the MTU, which then issues suitable commands to be executed on site. A system of either standard or customized software is used to collate, interpret, and manage the data.

SCADA as of now uses predominantly open-loop control systems and utilize predominantly long distance communications, though some closed-loop characteristics are often built in. As this is an open-loop system, it means that SCADA system cannot use feedback to check what results its inputs have produced. In other words, there is no machine-learning. However, the SCADA system allows a few operators to monitor the generation and transmission system and take action depending on overloading or out-of-limit voltage conditions.

1.2 Energy Control Center

Control center is the central nervous system of a power system. The development of control centers was prompted by the northeast United States blackout of 1965. It was then decided by the utilities to expand the effective use of computers in power system planning and operation. The resulting computer-based control centers, called the energy management systems, achieved a quantum jump in terms of intelligence and application software capabilities. The initial stage of conventional control centers were functioning with some or all of these operating characteristics:

- Analog metering with strip charts of key quantities;
- Analog generation control;
- Some manned substations;
- Supervisory control center at some central place or substation;
- Security analysis performed seasonally.

The present-day control centers are characterized by more centralization, use digital data, and perform regular online/offline security analyses. The functions of modern control centers are:

- Short medium and long term load forecasting
- System planning;
- Unit commitment and maintenance scheduling;
- Security monitoring;
- State estimation;
- Economic dispatch;
- Load frequency control.

1.3 Security Analysis and Monitoring

The most important concern for any electric utility is the security and stability of the power system. Power system security is the ability of the power system to withstand disturbances without interrupting the power supply and compromising the quality. In security applications the disturbances are more commonly called *contingencies.* The security assessment study is performed to check the vulnerability of the system against postulated contingencies on a real- or near real-time basis. A typical power system is never in steady state in the true sense because loads and generation patterns are continuously changing. Apart from this, the power system may be subjected to major disturbances such as equipment outage, loss of transmission line, or sudden change in large load. If any contingency occurs that results in violation of the tolerance limit of any equipment, it is automatically switched out of the system by protection devices. This event may be followed by a series of further actions that may switch other equipment out. Sometimes it may result in a cascading effect and ultimately the collapse of the system, resulting in a *blackout.*

Dy Liacco in 1967 identified that a power system may be operating in three possible states. These states are defined as the normal state, emergency state, and restorative state, as described earlier. In the emergency state, it is important to relieve the stress on the components that are working beyond their normal limits. Here the economics of operation is not relevant. In the restorative state, some portion of the system has lost power and the main objective is to restore the power as quickly as possible and bring the system back to the normal state.

There are three major components of power system security assessment [5]:

1. System monitoring;

2. Contingency analysis;
3. Preventive and corrective actions.

The most important aspect of security assessment is knowledge of the system states. Measurement devices are connected throughout the system to measure voltage, current, power flows, status of circuit breakers , switches, transformer taps, and so forth, and data is transmitted through telemetry to the control center. The computers in control centers can check the incoming information against prestored limits and send alarms to operators in the event of overload or violation of limits. These systems are combined with supervisory control systems that allow operators to control circuit breakers and disconnect switches and transformer taps from remote locations. These systems are known as the SCADA systems discussed earlier.

The second major security function is contingency analysis. In its simplest form a contingency analysis program models possible system troubles before they occur. We know that problems in a power system occur so quickly that the operators cannot act on them in time, which may cause serious damage to the system. Contingency analysis programs are based on a model of the power system and are used to study outage conditions and inform the operators to start corrective actions. Contingency analysis is the primary tool used for preparation of annual maintenance plans and the corresponding outage schedule for the power system.

1.4 State Estimation

Electric power system deals with the generation, transmission, and distribution of electric energy. The efficient and optimum economic operation and planning, along with security of electric power systems, have always occupied an important position in the power industry. In order to achieve these objectives, it is essential for power engineers to accurately monitor the power system operating states. An essential tool for monitoring the power system is state estimation. The vector consisting of bus voltage magnitudes and phase angles is called the state of a power system. In energy control centers, power system state estimation is carried out in order to provide best estimates of what is happening in the system based on real-time measurement and a predetermined system model. It is required in the critical operational functions of a power grid such as real-time security monitoring, load forecasting, economic dispatch, and load frequency control. The state estimator is a data processing scheme that computes the state of a system from the following three sources of information:

1. Measurement of system variables;
2. Mathematical model of the system;
3. Prior knowledge of system inputs and outputs known as pseudo-measurements.

Most of network applications use the real-time data provided by the state estimator. Therefore, an optimal performance of state estimation output is the ultimate concern for the system operator. This need is particularly more in focus today due to deregulated and congested systems and smart grid initiatives. The output of the state estimator nearly represents a true state of the system. However, discrepancies may occur due to incomplete measurements, meaning many variables are not measured or data is not available, inaccurate network parameters, and errors in measurements.

State estimators process the available measurements that include the line flows, injection measurements, and voltage measurements. The state vector is calculated considering the information from the network topology model and parameters using a predefined state estimation algorithm. If the state vector is obtained at an instant of time from the measurements also obtained at the same instant, it is known as static state estimation. In order to obtain the state vector regularly, the process of calculating the state vector is repeated at regular (a few seconds) intervals.

Before static state estimation techniques were used by energy control centers, operators were relying on the data available from power flows. However, when systems became large and the conditions and patterns of power flow became less predictable, an efficient and accurate state estimation became a prerequisite for the reliable operation of the power system.

Power system state estimation is one of the most important functions in a power system. It is required in the critical operational functions of a power grid such as real-time security monitoring, load forecasting, economic dispatch, and load frequency control. The static state estimation is related to conventional load flow calculations. The load flow calculations are performed on data that is free from errors. However, in state estimation the calculations are performed online using real-time measurements at different locations and transmitted to a central place. The data available at central station may have errors due to measurements, communication, and unexpected system changes. Thus in order to find error-free data, a large number of measurements are taken. This means that the number of input data, m, is more than the number of unknown variables, n. The static state estimator is a data processing algorithm used by the central computer to reduce total measurements and other information on an electric power system to estimate the static state vector.

The state estimation problem deals with various aspects of power system. An important feature of a state estimation program is to detect and eliminate bad data in a measurement set. Various techniques of static and dynamic state estimation, inclusion of current measurements, improvements in estimation with phasor measurement units, and distribution state estimation are all discussed in next few chapters.

References

[1] Dy Liacco, T. E., "Real Time Computer Control of Power Systems," *Proceedings of the IEEE,* Vol. 62, No. 7, 1974, pp. 884–891.

[2] Prince, B., et al., "Advances in Energy Management Systems," *IEEE Transactions on Power Systems,* Vol. 1, No. 3, 1986, pp. 226–234.

[3] Minakawa, T., "Schemes and Design for Advanced Energy Control Centers," *IEEE Transactions on Power Systems,* Vol. 6, No. 1, 1991, pp. 372–378.

[4] Hadbah, A., A. Kalam, and H. Al-Khalidi, "The Subsequent Security Problems Attributable to Increasing Interconnectivity of SCADA Systems," *Australian Universities Power Engineering Conference,* AUPEC 2008, Sydney, New South Wales, December 2008, pp. 1–4.

[5] Dy Liacco, T. E., "Modern Control Centers and Computer Networking," *IEEE Computer Applications in Power,* Vol. 7, No. 4, 1994, pp. 17–24.

2

Power Flow Equations

2.1 Power System Representation

Power flow problems, also known as load flow, are the heart of most power system planning studies. In order to study the operational features and electrical performance, the power system is assumed to be working in a symmetrical steady state. The power flow program computes the voltage magnitude and angle at each bus in a power system, under the balanced three-phase steady state condition. Three-phase load flows are used to analyze unbalanced three-phase systems (distribution systems). However, in this chapter the power system will be assumed to be working in a balanced steady state condition. Thus only the per-phase model is considered. From these values the real and reactive power flow in the system can be calculated. Before power flow equations can be formed and their solutions obtained, it is important to derive the mathematical models of the various components of the power system.

2.1.1 Transmission Lines

Transmission lines in a power system are classified according to their lengths as short, medium, or long lines. In general the short transmission lines have lengths less than 80 km and are represented by series impedance only. The typical length of medium transmission lines is from 80 to 250 km and are represented by lumped parameters in the form of nominal π network. Lengths of more than 250 km are classified as long transmission lines; the line parameters of these lines are assumed to be distributed uniformly throughout the length of the line. For steady state analysis the variables of interest are the voltages and the currents at the line terminals, V_S, I_S, V_R, and I_R. Here the subscripts S and R are used to signify the sending and receiving end, respectively. The sending end voltages and currents of the long transmission line are related to the receiving end voltage and current by ABCD parameters as

$$\begin{pmatrix} V_S \\ I_s \end{pmatrix} = \begin{bmatrix} \cos(h\gamma l) & Z_c \sin(h\gamma l) \\ \sin(h\gamma l) / Z_c & \cos(h\gamma l) \end{bmatrix} \begin{pmatrix} V_R \\ I_R \end{pmatrix} \tag{2.1}$$

where $Z_c = \sqrt{z/y}$ is the characteristic impedance of the line, and $\gamma = \sqrt{yz}$ is the propagation constant. As power system networks consist of a large number of lines, (2.1) is very convenient and a simpler way to represent each line by its π-equivalent shown in Figure 2.1. Here Z' and Y' are used.

$$Z' = Z_c \sinh(\gamma l) = \frac{Z \sin h(\gamma l)}{\gamma l} \tag{2.2}$$

$$\frac{Y'}{2} = \frac{\tanh(\gamma l/2)}{z_c} = \frac{Y/2 \tanh\left(\frac{\gamma l}{2}\right)}{\left(\frac{\gamma l}{2}\right)} \tag{2.3}$$

Example 2.1

A three-phase 500-kV transmission line is 300 km long. The series impedance is $z = 0.0165 + j\,0.3306\Omega$ per phase per km and the shunt impedance

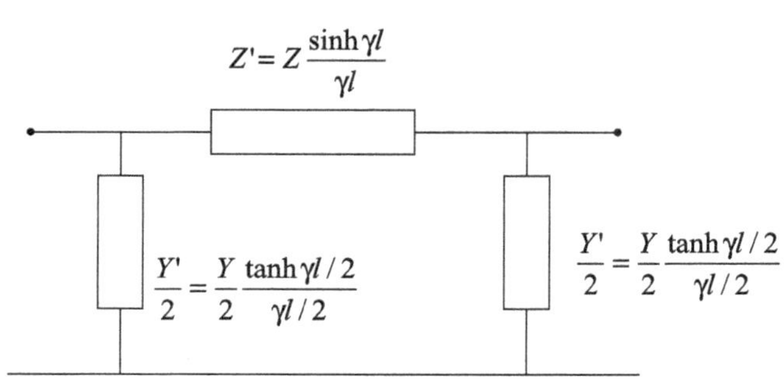

Figure 2.1 Equivalent model of a long transmission line.

is $y = j\,4.674 \times 10^{-6}$ siemens per phase per km. Determine the equivalent π model.

Solution

$$z = 0.0165 + j0.3306 = 0.3310 < 87.14$$

$$Z_c = \sqrt{\frac{z}{y}} = \sqrt{\frac{0.3310 < 87.10}{4.674 \times 10^{-6} < 90}} \quad 266.1 < -1.43$$

$$\gamma = \sqrt{yz} = \sqrt{(0.3310 < 87.1)4.674 \times 10^{-6} < 90} = \sqrt{1.547 \times 10^{-6}} < 177.14$$

$$\gamma l = \sqrt{1.547 \times 10^{-6}} < 177.14 \times 300$$

$$= 0.00931 + j0.3730 = 03731 < 88.57$$

$$\cosh(\gamma l) = 0.9313 + j0.0034 = 0.9313 < 0.209$$

$$\sinh(\gamma l) = 0.0087 + j0.3664 = 0.3645 < 88.63$$

$$Z' = Z_c \sinh(\gamma l) = 266.1 < -1.43 \times 0.345 < 88.63 = 91.80 < 87.17$$

$$\frac{Y'}{2} = \frac{\tanh\left(\gamma l/2\right)}{Z_c} = \frac{\cosh(\gamma l) - 1}{(\gamma l/2 \sinh(\gamma l) Z_c} = 7.095 \times 10^{-4} < 89.97$$

The equivalent model is shown in Figure 2.1.

2.1.2 Power Transformer

In a power system, three-phase power is transmitted over transmission lines and three-phase transformers are used to convert voltage from one level to another. The primary and secondary windings of three-phase transformers can be connected in either wye (Y) or delta (Δ) configurations. This results in four possible configurations: *Y-Y*, Y-Δ, Δ-Δ, Δ-Y as shown in Figure 2.2. The three-phase transformers are represented by their per-phase model. As well, in power system quantities such as voltage, current, power, and impedance are often represented in per-unit or percentage of specified base values. One advantage of using per-unit values is that by properly specifying base quantities, the transformer equivalent circuit can be simplified. Since the transformers are having different values of voltages and currents at the primary and secondary sides, by using per-unit values the voltages, currents, and impedances do not change when referred from one side of a transformer to the other side. This is a significant advantage in a power system where hundreds of transformers are involved. The per-unit quantities are calculated as follows:

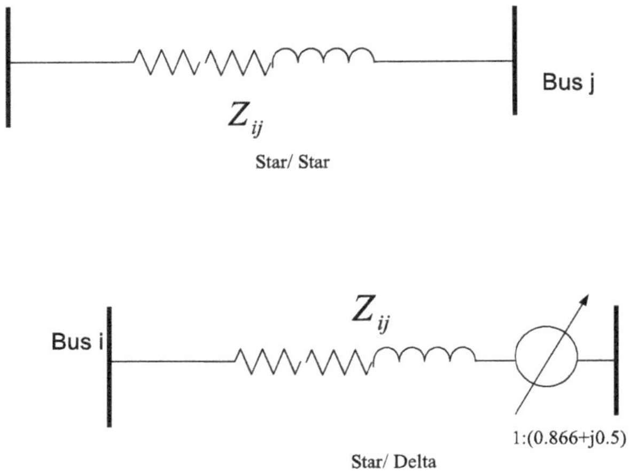

Figure 2.2 Star/star and star/delta transformers.

$$\text{Per-unit (p.u.) quantity} = \frac{\text{actual quantity}}{\text{base value}}$$

The base value has the same unit as the actual quantity, thus making the per-unit quantity dimensionless, and the base value is always a real number. The angle of the p.u. quantity is therefore same as the angle of actual quantity.

Two independent base values can be arbitrarily selected at one point in a power system. Normally base voltage V_{base} and power P_{base} are selected for one phase of a three-phase circuit. From these base quantities the base values current and impedance can be calculated as:

$$I_{base} = \frac{P_{base}}{V_{base}} \tag{2.4}$$

and

$$Z_{base} = R_{base} = X_{base} = \frac{V_{base}}{I_{base}} = \frac{V_{base}^2}{P_{base}} \tag{2.5}$$

Also by convention the value of the base power is selected to be the same for the entire power system and the ratio of voltage bases on either side of a transformer is selected to be the same as the ratio of the transformer voltage ratings. In this way the per-unit impedance remains unchanged when referred from one side of transformer to the other side.

For the three-phase system instead of obtaining the per-phase values, the per a unit values can be obtained directly by using three-phase base quantities.

Selecting three-phase MVA as base = MVA_{base}, line to line base voltage as KV_{base}.

For a star connection the base current is

$$I_{base} = 1000 \times \frac{MVA_{base}}{\sqrt{3}KV_{base}} A$$

and

$$\text{Base Impedance } Z_{base} = \frac{(KV_{base})^2}{MVA_{base}}$$

The transformers in power flow problems are represented by π network to make them compatible with transmission line models. Since the core losses and magnetization current for power transformers are of the order of 1% of maximum ratings, the shunt branch of transformer equivalent circuit is represented by only a leakage reactance and resistance representing winding and core losses for power flow calculations.

2.1.2.1 Fixed Tap Setting Transformer

Transformers with provision of tap changing are used to regulate the voltages at the bus. A transformer with a fixed tap setting is represented by its impedance Z_{ij} or admittance Y_{ij} in series with ideal transformer, if the transformer is connected at bus i in line i–j (Figure 2.3).

If the turn ratio of ideal transformer is a:

$$I_i = (V_i - aV_j)\frac{Y_{ij}}{a^2} = -I_j \tag{2.6}$$

The equivalent circuit of this transformer is shown in Figure 2.4.

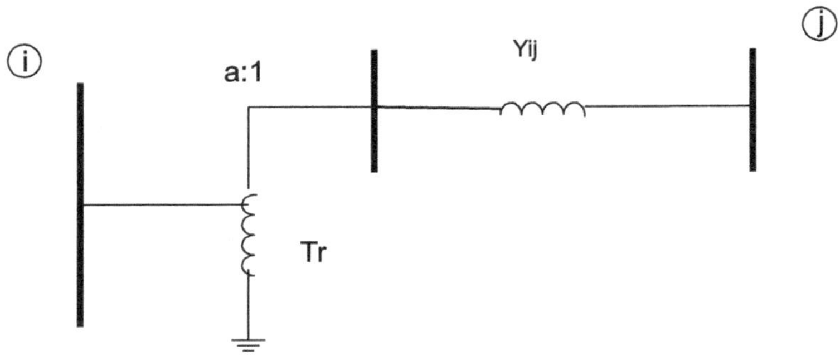

Figure 2.3 Circuit of transformer with fixed tap setting.

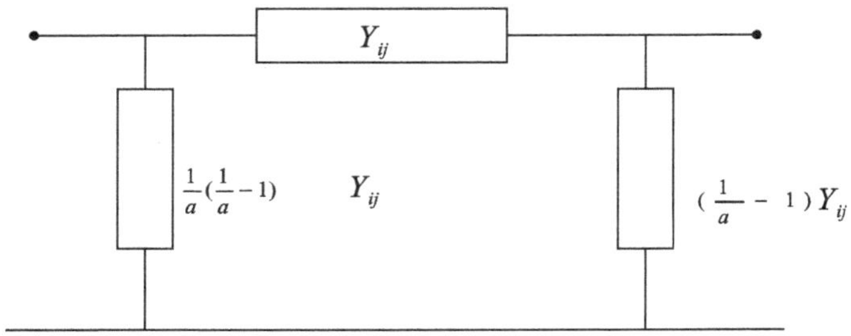

Figure 2.4 Equivalent π circuit of a transformer.

2.1.2.2 Phase-Shifting Transformer

A phase-shifting transformer is used to control the power flow over the line. It is represented by an ideal transformer with complex turns ratio ($a + jb$) in series with impedance or admittance. The circuit of this transformer is shown in Figure 2.5.

2.2 Admittance Diagram

For solving a power flow problem, the system is assumed to be operating under a balanced condition and is represented by a single-phase network. The network consists of hundreds of nodes and branches with impedances specified in p.u. on a common MVA base. Network equations for the solution of power flow problems are generally written in nodal admittance form. The formulation of network equations in nodal admittance forms results in complex linear simultaneous algebraic equations in terms of node currents.

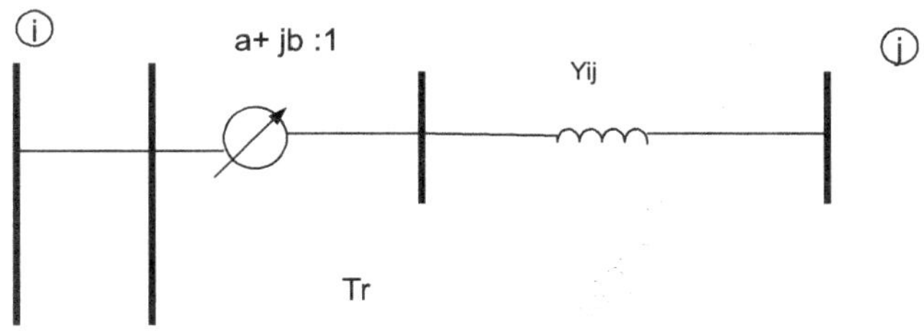

Figure 2.5 Equivalent circuit of a phase-shifting transformer.

In order to obtain the node voltage equation, bus admittance matrix is formulated first. The bus admittance comprises of driving point admittance for buses and the transfer admittances between buses. The formulation of Y_{bus} is demonstrated for a four-bus system in Example 2.2.

Example 2.2

A single line diagram of a four-bus power system is shown in Figure 2.6. The data is given in per unit on a common MVA base and resistances are neglected.

Solution

After converting the impedances into admittances, the admittance diagram of the power system is shown in Figure 2.7.

For solving the nodal equations node 0 (normally ground) is taken as reference. Applying KCL to nodes 1–4 we get:

$$I_1 = y_{11}V_1 + y_{12}(V_1 - V_2) + y_{13}(V_1 - V_3) \tag{2.7}$$

$$I_2 = y_{20}V_2 + y_{12}(V_2 - V_1) + y_{23}(V_2 - V_3) \tag{2.8}$$

$$0 = y_{23}(V_3 - V_2) + y_{13}(V_3 - V_1) + y_{34}(V_3 - V_4) \tag{2.9}$$

$$0 = y_{34}(V_4 - V_3) \tag{2.10}$$

or

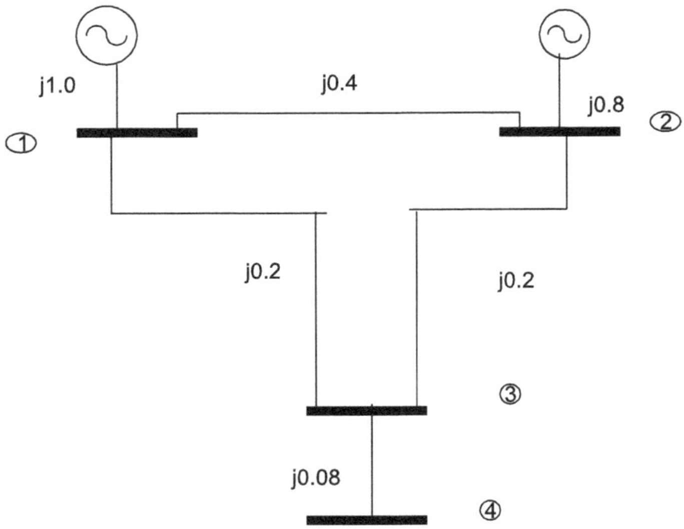

Figure 2.6 Impedance diagram of a four-bus system.

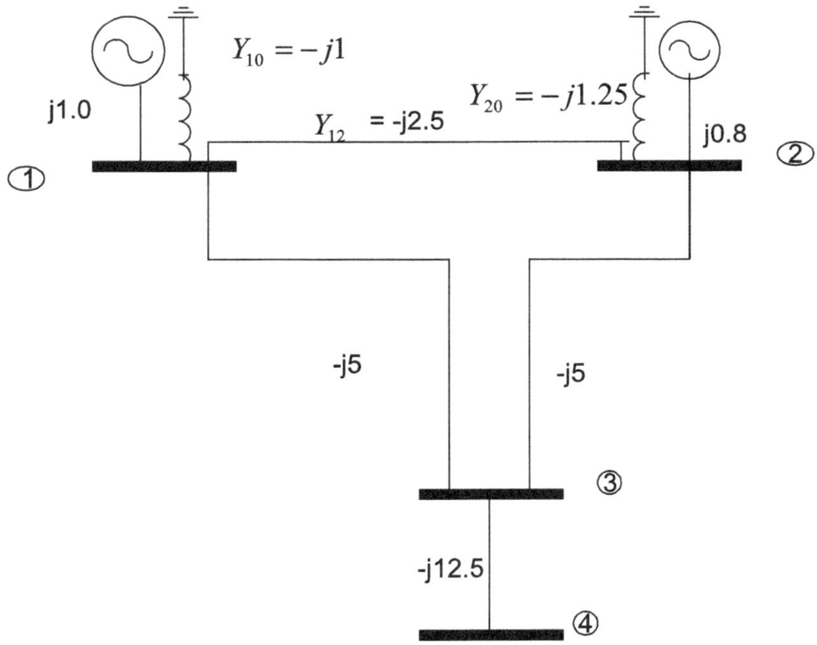

Figure 2.7 Admittance diagram of a four-bus system.

$$I_1 = \left(y_{10} + y_{12} + y_{13}\right)V_1 - y_{12}V_2 - y_{13}V_3 \tag{2.11}$$

$$I_2 = -y_{12}v_1 + \left(y_{20} + y_{12} + y_{23}\right)V_2 - y_{23}V_3 \tag{2.12}$$

$$0 = -y_{13}v_1 - y_{23}V_2 + \left(y_{13} + y_{23} + y_{34}\right)V_3 - y_{34}V_4 \tag{2.13}$$

In matrix form these equations can be written as:

$$\begin{bmatrix} I_1 \\ I_2 \\ I_3 \\ I_4 \end{bmatrix} = \begin{bmatrix} Y_{11} & Y_{12} & Y_{13} & Y_{14} \\ Y_{21} & Y_{22} & Y_{23} & Y_{24} \\ Y_{31} & Y_{32} & Y_{33} & Y_{34} \\ Y_{41} & Y_{42} & Y_{43} & Y_{44} \end{bmatrix} \begin{bmatrix} V_1 \\ V_2 \\ V_3 \\ V_4 \end{bmatrix} \tag{2.14}$$

where

$Y_{11} = y_{10} + y_{12} + y_{13}$
$Y_{22} = y_{20} + y_{12} + y_{23}$
$Y_{33} = y_{13} + y_{23} + y_{34}$
$Y_{44} = y_{34,}\ Y_{12} = Y_{21} = -\ y_{12},\ Y_{13} = \mathrm{Y}_{31} = -y_{13},\ Y_{23} = Y_{32} = -y_{23},\ Y_{34}$
$=$
$Y_{43} = -y_{34}$

If there are n buses in the system the node voltage equation can be written as:

$$\begin{bmatrix} I_1 \\ I_2 \\ \vdots \\ I_i \\ \vdots \\ I_n \end{bmatrix} = \begin{bmatrix} Y_{11} & Y_{12} & \cdots & Y_{1i} & \cdots & Y_{1n} \\ Y_{21} & Y_{22} & \cdots & Y_{2i} & \cdots & Y_{2n} \\ \vdots & \vdots & \vdots & \vdots & \vdots & \vdots \\ Y_{1i} & Y_{i2} & \cdots & Y_{ii} & \cdots & Yin \\ \vdots & \vdots & \vdots & \vdots & \vdots & \vdots \end{bmatrix} \begin{bmatrix} V_1 \\ V_2 \\ \vdots \\ V_i \\ \vdots \\ V_n \end{bmatrix} \tag{2.15}$$

or

$$\boldsymbol{I}_{bus} = \mathbf{Y}_{bus}\ \mathbf{V}_{bus} \tag{2.16}$$

where I_{bus} is the vector of currents injected into the bus. It is positive when entering the bus and is negative if flowing out of the bus. V_{bus} is the vector of bus voltages measured with respect to reference (ground) bus.

2.3 Power Flow Analysis

In solving power flow problem, four quantities associated with each bus are considered. These are voltage magnitude $|V|$, phase angle δ, real power P, and reactive power Q. Depending on the quantities specified at the bus these are classified as follows:

Slack bus: One of the buses is taken as the reference bus where the magnitude and phase angle of the voltage are specified. This bus is known as the slack bus or swing bus. It makes up the difference between the scheduled loads and generated power caused by the losses in the network.

Load bus: At these buses the active and reactive power are specified. The unknown quantities are the magnitude and angle of bus voltages. These buses are also known as P-Q buses.

Generation bus: Also known as voltage-controlled bus or P-V bus. Here the voltage magnitude V and the real power P are known. The unknown parameters are voltage phase angle and reactive power Q.

The basic equations for load flow studies are nothing but a set of network equations that are in the form of nonlinear algebraic equations. These equations can be formed either in the loop frame of reference or the bus frame of reference. Also either the admittance or the impedance bus matrix can be used. However, today the bus frame of reference both in the admittance form and impedance form is more commonly used. In an earlier section formulating a bus admittance matrix was demonstrated. The Y_{bus} matrix as shown is sparse (i.e., it contains nearly 85 percent elements as zero at the off-diagonal places). By using sparcity techniques a lot of computer space can be saved.

In solving the power flow problem using bus admittance formulation, only the positive sequence network is considered and any mutual coupling between the elements is neglected.

The performance equation for the power flow analysis in bus admittance form is (2.16).

In the following equation, in the ith bus the current and the real and reactive power is given by:

$$I_i = \sum_{j=1}^{n} Y_{ij} V_j \quad j \neq i \tag{2.17}$$

$$P_i + jQ_i = V_i I_i^* = V_i \sum_{j=1}^{n} Y_{ij} V_j \quad j \neq i \tag{2.18}$$

or

$$I_i = \frac{P_i - jQ_i}{V_i^*} = V_i \sum_{j=0}^{n} Y_{ij} - \sum_{j=1}^{n} Y_{ij} \tag{2.19}$$

Also

$$V_i = \frac{1}{Y_{ii}} \left(I_i - \sum_{j=1}^{n} Y_{ij} V_j \right) \quad j \neq i \tag{2.20}$$

These equations can also be written in the form:

$$P_i = |V_i| \sum_{j=1}^{n} |V_j| |Y_{ij}| \cos(\theta_{ij} + \delta_j - \delta_i) \tag{2.21}$$

$$Q_i = |V_i| \sum_{j=1}^{n} |V_j| |Y_{ij}| \sin(\theta_{ij} + \delta_j - \delta_i) \tag{2.22}$$

where $i = 1, 2, \ldots, n$.

These nonlinear algebraic equations can be solved by any iterative techniques.

2.3.1 Gauss-Seidel Method

The basic procedure for the Gauss-Seidel method of solving the nonlinear equations (2.16) and (2.18) is as follows:

We assume that all the buses except the slack bus are load buses only.

1. Calculate the $\mathbf{Y}_{bus}$ matrix for the network including the admittances of all transmission lines and transformers between buses but exclude the admittances of loads.

2. Select one of the buses as the slack bus.
3. Select initial estimates for all bus voltages: usually it is assumed as $V = |1.0|\ p.u.$ angle delta = 0 (flat start).
4. Write the voltage equation for every bus in the system as:

$$V_i = \frac{1}{Y_{ii}}\left[\frac{P_i - jQ_i}{V_i^*} - \sum_{j=1}^{n} Y_{ij}V_j\right] \tag{2.23}$$

5. Calculate an updated voltage values at each load bus in succession using (2.23). This equation is based on the fact that the current injected into the network must be equal to the current flowing along all lines.
6. Calculate the difference between the new and old values of voltages. If the differences are less than a specified value, then the process is stopped; otherwise Step 5 is repeated.
7. The solution is then checked for valid values. Normally the bus voltages should not be far off from 1.0, and the angle not more than 45 degrees.

2.3.1.1 Power Flow Analysis with Generator Buses Included

At a generator bus the real power P and the magnitude of a bus voltage is specified. However, the limit of reactive power Q_{min} and Q_{max} are also specified.Since reactive power at this bus is not known, (2.23) can be applied only after estimating Q. The reactive power at bus i can be estimated from (2.23) as:

$$\begin{aligned} V_i &= \frac{1}{Y_{ii}}\left[\frac{P_i - jQ_i}{V_i^*} - \sum_{j=1}^{n} Y_{ij}V_j\right] \quad \text{or} \\ P_i - jQ_i &= V_{i\ i}^{*}\left[Y_{ii}V_i - \sum_{j=1}^{n} Y_{ij}V_j\right] \quad j \neq i \end{aligned} \tag{2.24}$$

Bringing $j = i$ into consideration

$$P_i - jQ_i = V_i^*\left[\sum_{j=1}^{n} Y_{ij}V_j\right] \quad \text{or} \quad Q_i = -\operatorname{Im} V_i^*\left[\sum_{j=1}^{n} Y_{ij}V_j\right] \tag{2.25}$$

Once the reactive power at the bus is estimated, then (2.23) can be used to update the bus voltage. However, if the reactive power estimated is more than Q_{max} or less than Q_{min}, then substitute $Q = Q_{max}$ if $Q \geq Q_{max}$ and $Q = Q_{min}$ if Q is $\leq Q_{min}$. The voltage magnitude estimate of the bus is forced to remain constant by multiplying it by the ratio of the old estimate to new estimate.

Example 2.3

A simple power system with 5 buses and 6 transmission lines is shown in Figure 2.8. The input data is given in Tables 2.1 and 2.2.

Solution:

Bus 1 is selected as the slack bus. Bus 2 is the voltage-controlled bus. Buses 3, 4, and 5 are load buses.

The Y_{bus} is determined first.

$$y_{12} = {}^{1}\!/_{0.02 + j0.04} = 10 - j20$$
$$y_{23} = 0.9615 - j4.8077$$
$$y_{35} = 0.822 - j2.192$$
$$y_{34} = 5 - j15$$
$$y_{45} = 10 - j20$$
$$y_{15} = 1.724 - j4.310$$

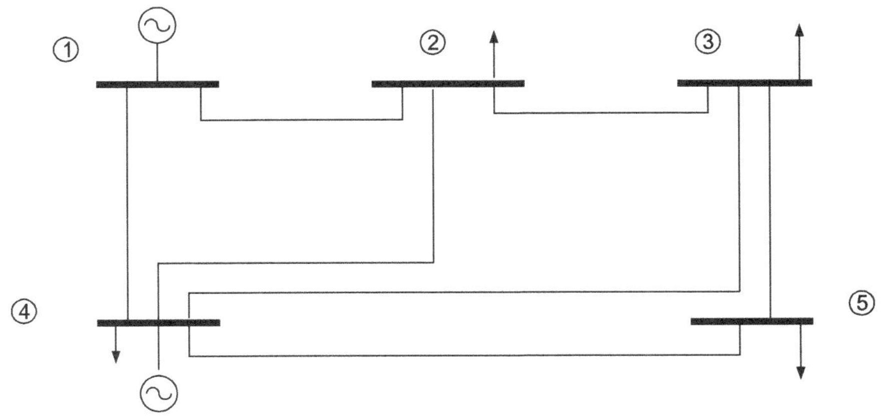

Figure 2.8 Single-line diagram of a power system.

Table 2.1
Line Input Data

Bus No.	Impedance	Line Charging
1–2	0.02+j0.04	J0.020
2–3	0.04+j0.2	J0.020
3–5	0.15+j0.4	J0.025
3–4	0.02+j0.06	J0.01
4–5	0.02+j0.04	J0.01
1–4	0.08+j0.2	J0.02

Table 2.2
Bus Input Data

Bus No.	Generation MW	Generation MVAr	Load MW	Load MVAr
1	0	0	0	0
2	50	25	15	10
3	0	0	45	20
4	0	0	40	15
5	0	0	50	25

and admittance to ground is:

$$y_1 = j0.02 + j0.02 = 0.004$$
$$= y_2 = 0.04, \;\; y_3 = j0.55, \;\; y_4 = j0.02, \; y_5 = j0.03$$

Elements of the bus admittance matrix are:

$$Y_{11} = y_{1+}y_{12} + y_{15} = 11.72 - j24.27$$

Similarly, other elements are found. Then Y_{bus} is calculated as:

$$Y_{bus} = \begin{bmatrix} 11.7241 \_ j24.2703 & -10 + j20 & 0 & 0 & -1.7241 + j4.3103 \\ -10 + j20 & 10.9615 - j24.7676 & -0.9615 + j4.8076 & 0 & 0 \\ 0 & -0.9615 + j4.8076 & 6.7834 - j21.9445 & -5 + j15 & -0.8219 + j2.1917 \\ 0 & 0 & -5 + j15 & 15 - j34.98 & -10 + j20 \\ -1.7241 + j4.3103 & 0 & -0.8219 + j2.1917 & -10 + j20 & 12.546 - j26.4471 \end{bmatrix}$$

Using the Gauss-Seidel method, the first estimate of voltages are found using Flat start and base MVA =100.00.

$$V_2^{(1)} = \frac{1}{Y_{22}}\left[\frac{P_2 - jQ_2}{V_2^{(0)*}}\right] - Y_{21}V_1 - Y_{23}V_3^{(0)}$$

$$V_2^{(1)} = \frac{1}{10.96 - j24.76}[0.35 - j0.15] -$$

$$[-10 + j20 - 0.9615 + j4.8076]$$

$$= \frac{0.35 - j0.15}{27.08 < -86.4} - [-10.9615 + j24.7676] =$$

$$11.3115 - j24.9576/27.08 < -86.4 = 1.01096$$

Next, the above value of $V_2^{(1)}$ is used to calculate the new value of $V_2^{(1)}$ as:

$$V_2^{(1)} = \frac{1}{Y_{22}}\left[\frac{P_2 - jQ_2}{V_2^{(1)*}} - Y_{21}V_1 - Y_{23}V_3^{(0)}\right]$$

The difference between the old value of voltage is compared with new voltage until the difference is less than 0.0004.

Using computer program the obtained result is shown in Table 2.3.

2.3.2 Newton-Raphson Method

The problem with the Gauss-Seidel method is that the convergence is very slow. As well; sometimes the convergence may not be obtained in spite of the fact that the solution exists. More efficient methods of solution for power flow are therefore needed with better convergence. One such method is the

Table 2.3
Solution of Example 2.3

Bus No.	Voltage Magnitude	Voltage Angle
1	1.0	0
2	0.9844	–0.4697
3	0.8813	–8.1707
4	0.8692	–8.8209
5	0.8750	–8.2003

Newton-Raphson (N-R) method. The N-R method for solution of nonlinear algebraic equation can be explained as described here.

Let us consider a set of n nonlinear equations written as:

$$\begin{aligned} f_1(x_1, x_2 \ldots\ldots\ldots x_n) &= n_1 \\ f_2(x_1, x_2 \ldots\ldots\ldots x_n) &= n_2 \\ f_n(x_1, x_2 \ldots\ldots\ldots x_n) &= n_n \end{aligned} \tag{2.26}$$

Then another set of equations can be written as:

$$\begin{aligned} g_1(x_1, x_2 \ldots\ldots\ldots x_n) &= f_1(x_1, x_2 \ldots\ldots\ldots x_n) - n_1 = 0 \\ g_2(x_1, x_2 \ldots\ldots\ldots x_n) &= f_2(x_1, x_2 \ldots\ldots\ldots x_n) - n_2 = 0 \\ g_n(x_1, x_2 \ldots\ldots\ldots x_n) &= f_n(x_1, x_2 \ldots\ldots\ldots x_n) - n_n = 0 \end{aligned} \tag{2.27}$$

Let us assume that the initial values of the variables x_1, x_2, …, x_n are $x_1^{(0)}, x_2^{(0)}, \ldots, x_n^{(0)}$. Now if the values are corrected by taking the values as:

$$\begin{aligned} x_1^{(1)} &= x_1^{(0)} + \Delta x_1^{(0)} \\ x_2^{(1)} &= x_2^{(0)} + \Delta x_2^{(0)} \\ x_n^{(1)} &= x_n^{(0)} + \Delta x_n^{(0)} \end{aligned} \tag{2.28}$$

Then the functions in (2.27) can be written as:

$$g_k\left(x_1^{(1)}, x_2^{(1)} \ldots\ldots\ldots x_n^{(1)}\right) \tag{2.29}$$

$$= g_k\left(x_1^{(0)} + \Delta x_1^{(0)}, x_2^{(0)} + \Delta x_2^{(0)} + \ldots + \ldots\ldots x_n^{(0)} + \Delta x_n^{(0)}\right) = 0 \tag{2.30}$$

if the nonlinear function $g(x)$ is initialized around the nominal values $x_1^{(0)}, x_2^{(0)}, \ldots, x_n^{(0)}$. Using the Taylor series and neglecting the higher-order terms, we then get:

$$g_k\left(x_1^{(1)}, x_2^{(1)}, \ldots, x_n^{(1)}\right) = g_k\left(x_1^{(0)}, x_2^{(0)}, \ldots, x_n^{(0)}\right) + \Delta x_1^{(0)} \frac{\partial g_k}{\partial_{x1}}\bigg|^0 + \Delta x_2^{(0)} \frac{\partial g_k}{x2}\bigg|^0 \tag{2.31}$$

where $\left.\dfrac{\partial g_k}{\partial x_i}\right|^0$ is the partial derivative of g_k evaluated at $x_2^{(1)} \ldots, x_n^{(1)}$.

Equation (2.29) can be written in vector matrix form as:

$$\begin{bmatrix} \partial g_1/\partial x_1 & \partial g_1/\partial x_2 & \cdots & \partial g_1/\partial x_n \\ \partial g_2/\partial x_1 & \partial g_2/\partial x_2 & \cdots & \partial g_2/\partial x_n \\ \vdots & \vdots & \vdots & \vdots \\ \partial g_n/\partial x_1 & \partial g_n/\partial x_1 & \cdots & \partial g_n/\partial x_n \end{bmatrix} \begin{bmatrix} \Delta x_1^{(0)} \\ \Delta x_2^{(0)} \\ \vdots \\ \Delta x_n^0 \end{bmatrix} = \begin{bmatrix} 0 - g_1(x_1^{(0)} \ldots x_n^{(0)}) \\ 0 - g_2(x_1^{(0)} \ldots x_n^{(0)}) \\ \vdots \\ 0 - g_n(x_1^{(0)} \ldots x_n^{(0)}) \end{bmatrix} \quad (2.32)$$

The solution of (2.32) can be written as:

$$\begin{bmatrix} \Delta x_1^{(0)} \\ \Delta x_2^{(0)} \\ \vdots \\ \Delta x_n^0 \end{bmatrix} = \left[J^{(0)}\right]^{-1} \begin{bmatrix} \Delta g_1^{(0)} \\ \Delta g_2^{(0)} \\ \vdots \\ \Delta g_n^{(0)} \end{bmatrix} \quad (2.33)$$

where the $[J]$ matrix is known as the Jacobian matrix. In addtion, $\Delta g_k^{(0)} = g_k\left(x_1^{(1)}, x_2^{(1)} \ldots\ldots\ldots x_n^{(1)}\right) - g_k\left(x_1^{(0)}, x_2^{(0)}, \ldots, x_n^{(0)}\right), k = 1 \ldots n.$

Since in making the simplification by neglecting higher-order terms of the Taylor series, (2.33) will not give the correct solution at the end of first iteration and we have to find the Jacobian $J^{(2)}$ until Δg_k becomes less than a specified small quantity ε.

2.3.2.1 Load Flow Algorithm Using the N-R Method

To apply the N-R method to load flow studies, let there be an n-bus power system, containing n_l number of load buses and n_g number of generator buses. Therefore $n = n_l + n_g + 1$. One bus is assumed to be slack bus. The state vector of X of the power system is:

$$X = \begin{bmatrix} \delta_1 \\ \delta_2 \\ \vdots \\ \delta_{n-1} \\ V_1 \\ V_2 \\ \vdots \\ V_{n-1} \end{bmatrix} = \begin{bmatrix} \delta \\ \boldsymbol{V} \end{bmatrix} \quad (2.34)$$

In (2.34) δ_i is the phase angle of the ith bus with respect to the slack bus, and V_i is the voltage magnitude of ith bus. The $2n - 2$ nonlinear equations **f**(X) are the calculated real and reactive power injections, and **b** is the specified power at the buses. The power flow problem is to find the power system state vector X^K which satisfies the desired power at the buses [3–5].

Or

$$\boldsymbol{f}\left(\boldsymbol{X}^{K}\right) = \boldsymbol{b} \quad (2.35)$$

The power injection at bus i is given by:

$$P_i + jQ_i = V_i I_i^* = V_i \sum_{j=1}^{n} Y_{ij} V_j \quad j \neq i \quad (2.36)$$

This with admittance formulation in rectangular coordinates can be written as:

$$V_i = e_i + jf_i, V_j = e_j + jf_j, Y_{ij} = G_{ij} + B_{ij} \quad (2.37)$$

Equation (2.36) therefore can be written as:

$$P_i + jQ_i = \left[\left(e_i + jf_{i,}\right) \sum_{j=1}^{n} \left(G_{ij} + B_{ij}\right)\left(e_j + jf_j\right) \right] \quad (2.38)$$

The real and reactive power at bus i is:

$$P_i = \sum_{j=1}^{n} e_i \left(e_j G_{ij} + f_j B_{ij}\right) + f_i \left(f_j G_{ij} - e_j B_{ij}\right) \quad (2.39)$$

$$Q_i = \sum_{j=1}^{n} f_i \left(e_j G_{ij} + f_j B_{ij} \right) - e_i \left(f_j G_{ij} - e_j B_{ij} \right) \tag{2.40}$$

The vector **b** consists of 2(*n*–1) specified injections and the difference between specified injection and the *k*th iteration calculated power is called the bus power mismatch. Then assuming bus 1 as the reference bus

$$\begin{bmatrix} \Delta P_2 \\ \vdots \\ \Delta P_n \\ \vdots \\ \Delta Q_1 \\ \vdots \\ \Delta Q_n \end{bmatrix} = \begin{bmatrix} J_{11} & J_{12} \\ J_{21} & J_{22} \end{bmatrix} \begin{bmatrix} \Delta e_2 \\ \vdots \\ \Delta n \\ \vdots \\ \Delta f_2 \\ \vdots \\ \Delta f_n \end{bmatrix} \tag{2.41}$$

where

$$[J_{11}] = \begin{bmatrix} \frac{\partial P_2}{\partial e_2} & \cdots & \frac{\partial P_2}{\partial e_n} \\ \vdots & \vdots & \vdots \\ \frac{\partial P_n}{\partial e_n} & \cdots & \frac{\partial P_n}{\partial e_n} \end{bmatrix} \tag{2.42}$$

$$[J_{12}] = \begin{bmatrix} \frac{\partial P_2}{\partial f_2} & \cdots & \frac{\partial P_2}{\partial f_n} \\ \vdots & \vdots & \vdots \\ \frac{\partial P_n}{\partial f_n} & \cdots & \frac{\partial P_n}{\partial f_n} \end{bmatrix} \tag{2.43}$$

$$[J_{21}] = \begin{bmatrix} \frac{\partial Q_2}{\partial e_2} & \cdots & \frac{\partial Q_2}{\partial e_n} \\ \vdots & \vdots & \vdots \\ \frac{\partial Q_n}{\partial e_n} & \cdots & \frac{\partial Q_n}{\partial e_n} \end{bmatrix} \tag{2.44}$$

$$[J_{22}] = \begin{bmatrix} \frac{\partial Q_2}{\partial f_2} & \cdots & \frac{\partial Q_2}{\partial e_n} \\ \vdots & \vdots & \vdots \\ \frac{\partial Q_n}{\partial f_n} & \cdots & \frac{\partial Q_n}{\partial f_n} \end{bmatrix} \tag{2.45}$$

Load Flow Algorithm

The solution to the load flow problem using N-R method is obtained as follows:

- Select the initial values for the bus voltage magnitudes and angles for all the busses except the slack bus for which $|V| = 1.0$ and $\delta = 0$. Normally the initial values chosen are equal to slack bus values (flat start).
- From the initial values calculate the real and reactive power for all the buses except the slack bus using (2.39) and (2.40).
- Since the active and reactive power at the buses is known, the error in power mismatch can be obtained for iteration using (2.41).
- Now the elements of the Jacobian matrix are obtained with the latest bus voltages and calculated power [(2.42) to (2.45)].
- Now (2.41) is solved to determine the voltage magnitude and angle corrections.
- From the corrections a new set of voltage estimates are obtained.
- Check if all the mismatches are below a certain small quantity. If yes, terminate the process; otherwise repeat the whole procedure.

Formation of Jacobian Matrix

The submatrices of the Jacobian matrix can be formulated in polar coordinates:

For any bus i the power equation is

$$P_i + jQ_i = \sum_{j=1}^{n} \left|V_i V_j Y_{ijj}\right| e^{j} \left(\theta_{ij} + \delta_j - \delta_i\right) \tag{2.46}$$

or

$$P_1 = \sum_{j=1}^{n} \left|V_i V_j Y_{ijj}\right| \cos(\theta_{ij} + \delta_j - \delta_i) \tag{2.47}$$

and

$$Q_i = \sum_{j=1}^{n} \left| V_i V_j Y_{ij\,j} \right| \sin\left(\theta_{ij} + \delta_j - \delta_i\right) \tag{2.48}$$

Now for the Jacobian submatrix J_{11}, the off-diagonal elements are given as:

$$\frac{\partial P_i}{\partial \delta_j} = -\left| V_i V_j Y_{ij} \right| \sin\left(\theta_{ij} + \delta_j - \delta_i\right) \tag{2.49}$$

and diagonal elements are:

$$\frac{\partial P_i}{\partial \delta_j} = \sum_{j=1}^{n} \left| V_i V_j Y_{ij\,j} \right| \sin(\theta_{ij} + \delta_j - \delta_i) \quad j \neq i \tag{2.50}$$

which is given by:

$$\frac{\partial P_i}{\partial \delta_i} = [-Q_i - V_i]^2 \tag{2.51}$$

J_{12} is given by

$$\frac{\partial P_i}{\partial V_j} = -\left| V_i V_j Y_{ij} \right| \cos\left(\theta_{ij} + \delta_j - \delta_i\right) \tag{2.52}$$

and

$$\frac{\partial P_i}{\partial V_i} \sum_{k=1}^{n} \left| V_i V_j Y_{ij\,j} \right| \cos(\theta_{ij} + \delta_j - \delta_i) + 2V_i^2\, G_{ii} \tag{2.53}$$

J_{21} is given by:

$$\frac{\partial Q_i}{\partial \delta_j} = -\left| V_i V_j Y_{ij} \right| \cos\left(\theta_{ij} + \delta_j - \delta_i\right) \tag{2.54}$$

and

$$\frac{\partial Q_i}{\partial \delta_i} = -\sum_{j=1}^{n} \left| V_i V_j Y_{ijj} \right| \cos(\theta_{ij} + \delta_j - \delta_i) \quad i \neq j = P_i - V_i^2 G_{ii} \tag{2.55}$$

and finally is given by:

$$\frac{\partial Q_i}{\partial V_j} = -\left| V_i V_j Y_{ij} \right| \sin\left(\theta_{ij} + \delta_j - \delta_i\right) \tag{2.56}$$

and

$$\frac{\partial Q_i}{\partial V_i} = \sum_{j=1}^{n} \left| V_i V_j Y_{ijj} \right| \sin(\theta_{ij} + \delta_j - \delta_i) - 2V_i^2 B_{ii} \tag{2.57}$$

Example 2.4

A five-bus power system is shown in Figure 2.8. The line impedance and line charging data is given in Table 2.4. The scheduled generation and loads are given in Table 2.5. Obtain the load flow solution by the N-R method using Y_{bus}.

Bus 1 is assumed to be the slack bus where active and reactive generation is unknown, but voltage magnitude is known =1.05 and angle is 0. Bus 5 is the PV bus where reactive power generated and voltage angle is unknown. Voltage magnitude is 1.02.

Table 2.4
Transmission Line Data

Bus No.	Impedance	Line Charging
1–2	0.02+j0.10	J0.030
1–5	0.05+j0.25	J0.020
2–3	0.04+j0.20	J0.025
2–5	0.05+j0.25	J0.020
3–4	0.05+j0.25	J0.020
3–5	0.08+j0.40	J0.010
4–5	0.10+j0.50	J0.075

Table 2.5
Bus Data

Bus No.	Generation MW	Generation MVAR	Load MW	Load MVAR
1	–	—	0	0
2	0	0	96	62
3	0	0	35	14
4	0	0	16	8
5	48	—	24	11

Solution

To begin with, voltages at bus 2, 3 and 4 are assumed 1.0 p.u. The angles are taken as 0= equal to slack bus angle.

The Y_{bus} matrix of the system shown in Figure 2.8 is obtained as shown in Table 2.6.

Using the polar method, the off-diagonal elements of $j_{11} = \frac{\partial P_i}{\partial \ddot{a}_j} = -\left|V_i V_j Y_{ij}\right| \sin\left(\theta_{ij} + \delta_j - \delta_i\right)$ and diagonal elements are obtained from $\frac{\partial P_i}{\partial \delta_i} = j \neq i$. The diagonal element of first row is $\frac{\partial P_i}{\partial \delta_i} = -Q_i - \left|V_i\right|^2 B_{ii}$.

The Jacobians are obtained as:

$$J_{11}^{(0)} \begin{bmatrix} 18.8269 & -4.8077 & 0 & -3.9231 \\ -4.807 & 11.058 & -3.8462 & -2.4519 \\ 0 & -3.8462 & 5.8077 & -1.9615 \\ -3.9231 & -2.4519 & -1.9615 & 12.4558 \end{bmatrix}$$

Table 2.6
Y_{bus} Matrix

2.6923–j13.4115	1.9231+j9.6154	0	0	–0.7692+j3.8462
1.9231+j9.6154	3.6538-j18.1942	–0.9615+j4.8077	0	–0.7692+j3.8462
0	–0.9615+j4.8077	2.2115–j11.0027	–0.7692+j3.8462	–0.4808+j2.4038
0	0	–0.7692+j3.8462	1.1538-j5.6742	–0.3846+j1.9231
–0.7692+j3.8462	–0.7692+j3.8462	–0.4808+j2.4038	–0.3846+j1.9231	2.4308-j11.8942

$$J_{12}^{(0)}\begin{bmatrix} 3.5423 & -0.9515 & 0 \\ -0.9615 & 2.2019 & -0.7692 \\ 0 & -0.7692 & 1.1462 \\ 0.7546 & -0.4904 & -0.3923 \end{bmatrix}$$

$$J_{21}^{(0)}\begin{bmatrix} -3.7654 & 0.9615 & 0 & 0.7846 \\ 0.9615 & -2.2212 & 0.7692 & 0.4904 \\ 0 & 0.7692 & -1.1615 & 0.3923 \end{bmatrix}$$

$$J_{22}^{(0)}\begin{bmatrix} 17.5615 & -4.8077 & 0 \\ -4.8077 & 10.8996 & -3.8462 \\ 0 & -3.8462 & -5.5408 \end{bmatrix}$$

From initial conditions the real and reactive power is calculated as:

$$P_{cal}^{(0)} = \begin{bmatrix} -0.1115 & -0.0096 & -0.0077 & -0.0098 \end{bmatrix}^T$$
$$Q_{cal}^{(0)} = \begin{bmatrix} -0.6327 & -0.1031 & -0.1335 \end{bmatrix}^T$$

The difference in power is obtained as:

$$\Delta P = \begin{bmatrix} -0.8485 & -0.3404 & -0.1523 & 0.2302 \end{bmatrix}^T$$
$$\Delta Q = \begin{bmatrix} 0.0127 & -0.0369 & 0.0535 \end{bmatrix}^T$$

The updates after the first iteration are:

$$\begin{bmatrix} V_2 \\ V_3 \\ V_4 \end{bmatrix} \text{magnitude} = \begin{bmatrix} 0.9864 \\ 0.9817 \\ 0.9913 \end{bmatrix}$$

and angle in degrees are:

$$\begin{bmatrix} \delta_2 \\ \delta_3 \\ \delta_4 \\ \delta_5 \end{bmatrix} = \begin{bmatrix} -4.91 \\ -6.95 \\ -7.19 \\ -3.09 \end{bmatrix}$$

After seven iterations, the load flow solution is obtained as shown in Table 2.7.

2.4 Decoupled Power Flow

The computational complexity of matrix inversion is the main limiting factor in finding a faster solution of the power flow problem. Numerical methods can be efficiently used if the advantage of physical properties of the system is taken into consideration. For example the sparsity in the Y_{bus} matrix can be exploited by using an ordered elimination technique. Another property of the system that can result in faster computation is by making a simplification of decoupling the real part of the Jacobian from the reactive part. This approximation is justified because the real power in a power system depends mainly on the angle difference, and the reactive power flow depends mainly on the voltage magnitude difference. The computational complexity of matrix inversion in inverting one matrix of size n is much more in terms of time as compared to inversion of two matrices of size $n/2$. The N-R method in polar coordinates can be written as:

Table 2.7
Load Flow Solution for Example 2.4

Bus No.	Voltage Mag.	Voltage Angle (deg)	Power Gen	Reactive Power Gen	Load MW	Load MVAR
1	1.05	0	126.6	57.11	0	0
2	0.9826	–5,0124	0	0	96	62
3	0.9777	–7.1322	0	0	35	14
4	0.9876	–7.3705	0	0	16	8
5	1.02	–3.2014	48	15.59	24	11

$$\begin{bmatrix} \Delta P_2 \\ \vdots \\ \Delta P_n \\ \vdots \\ \Delta Q_1 \\ \vdots \\ \Delta Q_n \end{bmatrix} = \begin{bmatrix} J_{11} & J_{12} \\ J_{21} & J_{22} \end{bmatrix} \begin{bmatrix} \Delta e_2 \\ \vdots \\ \Delta n \\ \vdots \\ \Delta f_2 \\ \vdots \\ \Delta f_n \end{bmatrix} = \begin{bmatrix} \Delta P \\ \Delta Q \end{bmatrix} = \begin{bmatrix} J_{11} & J_{12} \\ J_{21} & J_{22} \end{bmatrix} \begin{bmatrix} \Delta\delta \\ \dfrac{\Delta|V|}{V} \end{bmatrix} \tag{2.58}$$

If the approximation of decoupling is used (2.58) will become:

$$\begin{bmatrix} \Delta P \\ \Delta Q \end{bmatrix} = \begin{bmatrix} J_{11} & 0 \\ 0 & J_{22} \end{bmatrix} \begin{bmatrix} \Delta\delta \\ \dfrac{\Delta|V|}{V} \end{bmatrix} \tag{2.59}$$

The off-diagonal elements of the Jacobian can be written as:

$$\frac{\partial P_i}{\partial \delta_j} = -\left|V_i V_j Y_{ij}\right| \sin\left(\theta_{ij} + \delta_j - \delta_i\right)$$

which can be simplified as:

$$-\left|V_i V_j\right| s\left[B_{ij} \cos\left(\delta_j - \delta_i\right)\right] + \sin\left(\delta_j - \delta_i\right) = \frac{\partial Q_i \left|V_i\right|}{\partial V_j} \tag{2.60}$$

and diagonal elements are:

$$\frac{\partial P_i}{\partial \delta_i} = -B_{ii} V_i^2 - Q_i \tag{2.61}$$

and

$$\frac{\partial Q_i \left|V_i\right|}{\partial V_i} = B_{ii} V_i^2 + Q_i \tag{2.62}$$

Equations $\Delta P = J_{11}{}^{\Delta}$ and $\Delta Q = J_{22} \dfrac{\Delta|V|}{V}$ may be solved alternatively as decoupled Newton method. However, further simplifications can be made for a practical power system as the difference between phase angles $[(\delta_i - \delta_j)]$ is very small, or we can write:

$\cos(\delta_i - \delta_j) \approx 1,\ G_{ij}\sin(\delta_i - \delta_j) \ll B_{ij}$, and $Q_i \ll B_{ii}V_i^2$

With these approximations the elements of the Jacobian becomes:

$$\frac{\partial P_i}{\partial \delta_j} = \frac{\partial Q_i |V_i|}{\partial V_j} = -|V_i||V_j| B_{ij} \text{ for } i \neq j \tag{2.63}$$

and

$$\frac{\partial P_i}{\partial \delta_i} = -B_{ii}|V_i^2| = \frac{\partial Q_i |V_i|}{\partial V_i} \tag{2.64}$$

Substituting these in (2.64) we get:

$$[\Delta P_i] = |V_i||V_j|[B'_{ij}][\Delta\delta_j] \tag{2.65}$$

$$[\Delta Q_i] = |V_i||V_j|[B''_{ij}]\left[\Delta \frac{|V_j|}{V_j}\right] \tag{2.66}$$

Where B' and B'' are the elements of the $[-B]$ matrix.

2.5 Visual Tools for Power Flow Studies

The application of information visualization holds tremendous promise for the electric power industry. The power industry is still away from making full use of the potential available with visual representation. Prior work on visualizing electric power network has been limited to depicting raw or processed information on top of geographic layout. Recently a group at Pacific Northwest National laboratory, have developed an information visualization system prototype known as green grid that can be used to explore planning and monitoring of the American grid. A major challenge of green grid is to help in identifying the vulnerabilities of power grids by detecting the onset of network separation.

The development of a green grid involves a number of power grid visualization, graph drawing, and graph analytic topics. Historically AREVA's Energy Management System (EMS) and Power World Simulator are the two most common visualizing tools used by the power industry. However, their main features are to depict collected or processed information using color or icons on top of a geographic layout. The green grid graph layout is not a simple geographic or graphic template that ties icons or glyphs together. Rather, it has a visualization status that physically characterizes the contents and structurally characterizes the behavior of the underlying data for different reasoning purposes or analytical applications.

The visualization designers within the power industry have placed more attention on presenting the geographical information without giving much importance to the metadata of the graph in the layout. Green grid is developed that uses a weighted graph layout approach to model the physics of the power grid. The weighted graph approach to analyze the transmission of electric power along the grid represents a significant innovation in applying visualization in the electric power system. The ability to change the values of weights leads to the possibility of generating hundreds of plots from a single power flow or state estimation calculation.

For details about green grid software and its application, the reader is advised to read reference [6].

2.6 DC Power Flow

The term *dc power flow* comes from the earlier used network analyzer in which each network branch was represented by a dc resistance equivalent to its series reactance, and each current was proportional power flow in line. The dc modeling process starts from a simple transmission line connecting two nodes and is shown in Figure 2.9. The power flow equations can be written as [7]:

Sending end power

$$P_{ij} = gV_i(V_i - V_j \cos\delta) - V_iV_j b \sin\delta \tag{2.67}$$

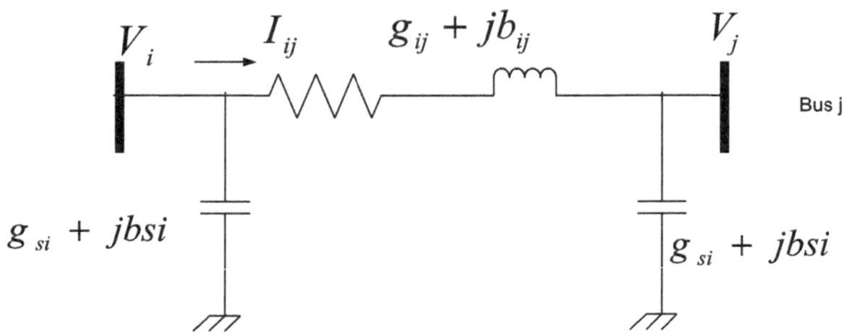

Figure 2.9 A π model of a line.

and receiving end power

$$P_{ji} = -gV_j\left(V_j - V_i\cos\delta\right) - V_iV_j b\sin\delta \tag{2.68}$$

A typical dc power flow model uses the following simplifications.

Losses are assumed to be zero or sending end power is assumed to be equal to the receiving end power.

$$P_{ij} = P_{ji} = V_j V_i b\sin\delta \tag{2.69}$$

$\sin\delta \cong \delta$, $V_i = V_j \cong 1.0$ and $b \cong \frac{1}{x}$

The power flow can be simplified $P \cong \delta/x$

This is the simplest model. However, for large power system studies it is essential to include the line losses. If line losses are not included for the networks where the X/R ratio is less than 4 times, the results will not be correct. As well, since the voltages are assumed to be 1.0 p.u. the errors will be large if the voltage variation is high.

If line losses are included then the equation for power can be written as:

$$P \cong h\delta \tag{2.70}$$

and sending end power $P_1 = P + \alpha_1$ and receiving end power $P_2 = P - \alpha_2$, where $h = 1/x$ and α_1 and α_2 are fixed injections whose sum approximates the line loss at some chosen operating point.

2.7 Regulating Transformers

Regulating transformers are used in a power system to control power flow and maintain voltage profile at a certain level. These transformers have two main units to perform two basic functions. The first unit is for basic phase shift to control flow of power, and the second unit is to control reactive power and voltage profile. Phase shifting is accomplished by generating a quadrature component of voltage which is added to the original voltage. An on-load tap changer is used to adjust the quadrature component for specific phase shift. Both components of voltages are injected in the line through series transformer. The most common arrangement uses a two winding star connected transformer for excitation and a delta connected transformer for injection. A regulating transformer is shown in Figure 2.8.

The presence of regulating transformers in lines modifies the Y_{bus}, thus modifying the power flow [8].

References

[1] Stevenson, W. D. Jr., *Elements of Power System Analysis,* McGraw Hill, 1982.

[2] Stagg, G. W., and A. H. El-Abiad, *Computer Methods in Power System Analysis,* McGraw Hill, 1968.

[3] Tinney, W. F., and C. E. Hart, "Power Flow Solution by Newton's Method," *IEEE Transactions on Power Apparatus and Systems,* PAS-86, 1967, pp. 1449–1556.

[4] Stott, B., and O. Alsac, "Fast Decoupled Load Flow," *IEEE Transactions on Power Apparatus System,* PAS-93, 1974, pp. 859–869.

[5] Monticelli, A., A. Garcia, and O. R. Savedra, "Fast Decoupled Load Flow: Hypothesis, Derivations, and Testing," *IEEE Transactions on Power Systems,* Vol. 5, 1990, pp. 1425–1431.

[6] Wong, P. C., et al., "A Novel Visualizing Technique for Electric Power Grid Analysis," *IEEE Transactions on Visualization and Graphics,* Vol. 15, No. 3, 2009, pp. 410-422.

[7] Benjamin, N., "Regular Transformer Model for use in Load Flow Studies," *IEEE Transactions on Power Apparatus and Systems,* Vol. 104, 1985, pp. 1102–1108.

3

Weighted Least Square Estimation

3.1 Introduction

A power system essentially consists of generation, transmission, and distribution systems. A transmission system contains a large number of substations that are interconnected by transmission lines, transformers, and other switching devices for system control and protection. The efficient and optimum economic operation and planning along with security of electric power systems have always occupied an important position in the power industry. In order to achieve these objectives, it is essential for power engineers to accurately monitor the power system operating states. An essential tool for real time monitoring of the power system is state estimation (SE). It determines the best estimates of the actual power system state based on available supervisory control and data acquisition (SCADA) measurements, power system model and other data. The weighted least square (WLS) estimation method is the most commonly used technique in state estimation of power system.

The idea of least square estimation has been used since the early part of the nineteenth century. The linear least square problem deals with an overdetermined system of linear equations (a system with more known equations than unknown). In a power system, the state variables are voltage magnitudes and relative phase angles at the system nodes. Static state estimation refers to the procedure of obtaining voltage magnitude and phase angles at all the nodes of a power system. The earliest application of state estimation in power system was given by Schweppe, et al. [1, 2] in the 1970s. Simultaneous measurements of data are required at all the nodes, and collection of this data at the energy control centers is used to determine the state of the system. However, such an approach is expensive and also vulnerable to errors in measurement and communication. The state estimation is therefore performed using statistical analysis on a set of m imperfect data telemeter from different nodes of a power system. In determining the state of power system the system is assumed to be balanced and operating at steady state. With these assumptions the power system can be represented by single-phase quantities of positive sequence components of the system devices. Also, as is the case of power flow analysis, all the network data and network variables are expressed in per-unit quantities. A number of books and technical papers have appeared in various journals on the topic of state estimation [3–9].

3.2 Properties of Weighted Least Square Estimates

The state estimation refers to a procedure where one uses samples of a large number of measurements, some of which may be inaccurate, to determine the best estimates of the states. Statistics can be helpful in manipulating the data according to known procedures used by statisticians. Of the many criteria that have been found useful, the following three are more suitable for state estimation in a power system:

1. *The maximum likelihood criterion*: Here the objective is to maximize the probability that the estimate of the state variable $\hat{x}$ is the true value of the state vector $\mathbf{x}$.
2. *The weighted least square criterion:* Here the objective is to minimize the sum of squares of the weighted deviations of the estimated measurements $\hat{z}$ from the actual measurements $\mathbf{z}$.
3. *The minimum variance criterion*: Here the objective is to minimize the expected value of the sum of the squares of the deviations of the estimated components of the state variable vector from the corresponding components of the true state vector.

3.3 Maximum Likelihood Weighted Least Square State Estimation

In the maximum likelihood state estimation method, the random measurement errors are assumed to have known probability distribution with unknown parameters. The probability density function for all the measurements can then be written in terms of these unknown parameters. This function is called the likelihood function. The estimation of the state is selected based on the method that maximizes this probability. The likelihood function will be maximum when the unknown parameters are closest to their actual values. Some of the measured quantities as obtained from various measuring devices are assumed to have errors. Thus the measured quantity differs from its actual value by an unknown random error.

$$z_{meas} = z_{act} + \eta \tag{3.1}$$

If the network is in steady state and the measurements performed a large number of times, the error would average to zero and $z_{meas} = z_{act}$. The measurement errors are normally assumed to have a Gaussian distribution. The parameters for Gaussian distribution are the mean μ and its variance σ^2. The problem of maximum likelihood estimation is therefore solved for these parameters. The probability density function of number z is assumed to have normal Gaussian distribution.

The Gaussian probability density function for the random number z is given by

$$f(z) = \frac{1}{\sigma\sqrt{2\pi}} \exp - {(z-\mu)^2} / {2\sigma^2} \tag{3.2}$$

Where σ is called the standard deviation, σ^2 is called the variance of the random number, and μ is mean or expected value of z. The f(**z**) describes the behavior of the random number. If $(z-\mu)/\sigma = \eta$, then the plot of f(η) versus σ, which is known as Gaussian probability density function and is shown in Figure 3.1. The value of σ gives an indication of random measurement error. If σ is large the measurement is inaccurate (poor measuring instrument), whereas a small value of σ indicates the good quality measuring device. The normal distribution is commonly used for modeling measurement errors as these errors are caused by many factors.

3.3.1 Likelihood Function

The likelihood function for a sample of m independently and identically distributed observations can be obtained by finding the joint probability density function (pdf). If there are m independent measurements and each has the same Gaussian pdf, and each measurement is assumed to be independent of

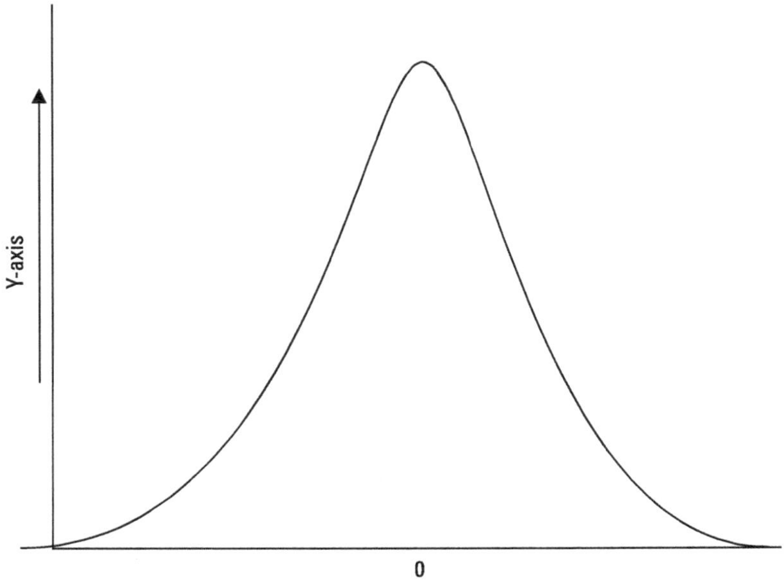

Figure 3.1 The normal (Gaussian) distribution curve.

the others, then the joint pdf can be simply written as the product of individual pdfs.

$$f_m(z) = f(z_1) f(z_2) \ldots f(z_m) \tag{3.3}$$

The function $f_m(z)$ is called the likelihood function for z. It is essentially a measure of probability of observing the particular set of measurements in the vector **z**. The maximum likelihood estimation of $f_m(z)$ is to maximize it by varying the assumed parameters μ and σ of the density function. A convenient transformation can be made at this point by maximizing the logarithm of the likelihood function. The logarithmic value of function $f(z)$ of (3.1) is given by

$$\log f(z) = -\frac{1}{2}\frac{(z-\mu)^2}{\sigma} - \log \sigma\sqrt{2\pi} \tag{3.4}$$

or the modified function $L = \log f_m(z)$.

$$\sum_{i=1}^{m} \log f(z_i) = \frac{1}{2}\sum_{i=1}^{m}\left(\frac{z_i - \mu_i}{\sigma_i}\right)^2 - \frac{m}{2}\log 2\pi - \sum_{i=1}^{m}\sigma_i \tag{3.5}$$

The maximum likelihood estimate is obtained by finding the maximum of $\log f_m(z)$, which can also be obtained by finding the minimum of

$$\min J(z) = \sum_{i=1}^{m}\left(\frac{z_i - \mu_i}{\sigma_i}\right)^2 \tag{3.6}$$

Equation (3.6) is known as *weighted least squares* estimator. It is equivalent to a maximum likelihood estimator if the measurement errors are represented by random numbers having a normal distribution.

The minimization problem can also be written in the form of measurement residuals. If r_i is residual of measurement i, then

$$r_i = z_i - \mu_i \tag{3.7}$$

Where the mean μ_i, which is the expected value of z_i, can be expressed as $h_i(x)$, a nonlinear function relating the system state vector **x** to the ith measurement. The square of residual r_i weighted by W_i is equal to inverse of error variance σ^2. Hence the minimization problem of (3.7) can also be written as

$$\text{minimize} \quad \sum_{i=1}^{m} W_i r_i^2 \tag{3.8}$$

subject to

$$z_i = h_i(x) + r_i \tag{3.9}$$

3.4 Matrix Formulation and Measurement Model

If we have a sample of m measurements given by vector **z**, and n states (voltage magnitude and angle), $n < m$, and e is m dimensional vector of measurement errors. Here h is the m dimensional nonlinear function vector relating measurements to the state vector. Then the measurement vector can be written in the matrix form as

$$z = \begin{bmatrix} z_1 \\ z_2 \\ \vdots \\ z_m \end{bmatrix} = \begin{bmatrix} h_1(x_1, x_2, \ldots, x_n) \\ h_2(x_1, x_2, \ldots, x_n) \\ \vdots \\ h_m(x_1, x_2, \ldots, x_n) \end{bmatrix} + \begin{bmatrix} e_1 \\ e_2 \\ \vdots \\ e_m \end{bmatrix} = h(x) + e \quad (3.10)$$

The weighted least square estimator will minimize the following objective function:

$$\min J(x) = \sum_{i=1}^{m} \left(z_i - h_i(x)\right)^2 / R_{ii} \quad (3.11)$$

where

$$R = \begin{bmatrix} \sigma_1^2 & & & \\ & \sigma_2^2 & & \\ & & \ddots & \\ & & & \sigma_m^2 \end{bmatrix} \quad (3.12)$$

where $[R]$ is called the *covariance matrix* of measurement errors.

Equation (3.11) can be written as

$$\min J(x) = [z - h(x)]^T [R^{-1}][z - h(x)] \quad (3.13)$$

The minimum of $J(x)$ can be obtained when $g(x) = \delta j(x)/\delta x = 0$

Or

$$-H^T(x) R^{-1}[z - h(x)] = 0 \quad (3.14)$$

where $H(x) = \delta h(x)/\delta x$ is the Jacobian matrix of dimension $(m \times n)$ measurement.

Note that (3.14) is valid only when $m > n$; that is, the number of parameters to be estimated is less than the number of measurements made. If $n > m$ there will be many values of states that will satisfy the above equation. In such cases, pseudomeasurements are added to give a completely determined or overdetermined problem.

Pseudomeasurements are not obtained from meters but are typically calculated using historical data or short-term load forecast, or as zero injections.

The most common exact pseudomeasurement is the bus injection at a substation that has no generation and serves no load. Here we can say with certainty that real and reactive power injection at this bus is zero.

The nonlinear function $h(x)$ can be linearized as

$$h(x+\Delta x) \approx h(x) + H(x)\Delta x \tag{3.15}$$

If we write $H(x)$ as H, the following iterative procedure is obtained:

$$\left(H^T R^{-1} H\right)\Delta x = H^T R^{-1}\left[z - h(x)\right] \tag{3.16}$$

Or

$$\Delta x = \left(H^T R^{-1} H\right)^{-1} H^T R^{-1}\left[z - h(x)\right] \tag{3.17}$$

where $x^{k+1} = x^k + \Delta x$.

$$x^{k+1} = x^k + (H\left(x^k\right)^T R^{-1} H\left(x^k\right)^{-1} H\left(x^k\right)^T R^{-1}\left[z - h\left(x^k\right)\right] \tag{3.18}$$

The symmetric matrix $(H^T R^{-1} H) = G(x)$ is called the *gain* or *information matrix* and (3.16) through (3.18) are known as normal equations.

3.4.1 Measurement Model

The state estimator needs a set of analog measurements and system topology to estimate the system states. The measurements can be expressed in terms of state variables either using the rectangular or polar coordinates. If there is N number of buses and polar coordinates are used, then the state vector will have $(2N\text{-}1)$ elements, N bus voltage magnitudes, and $(N\text{-}1)$ phase angles. In a power system the most common type of measurements are the power line flows, bus power injections, bus voltage magnitudes, and line current flow magnitudes. A necessary requirement for the number of measurements is that the system is observable. This is possible only if the number of available measurements is equal or larger than the number of estimated variables. The power flow measurements and power injection measurements are directly expressed in terms of corresponding state variables as follows: the equivalent circuit of line is shown in Figure 3.2.

The real and reactive power injection at bus i is expressed as

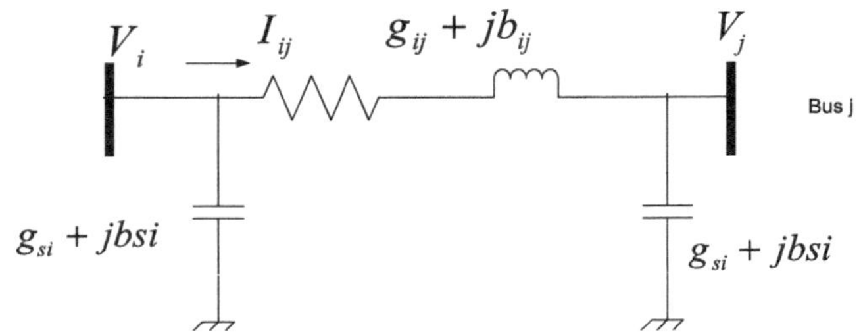

Figure 3.2 A π model of line.

$$P_i = V_i \sum_{j=1}^{N} V_j \left[G_{ij} \cos(\theta_i - \theta_j) + B_{ij} \sin(\theta_i - \theta_j) \right] \quad j = 1,2 \ldots N, \quad j \neq i \quad (3.19)$$

$$Q_i = V_i \sum_{j=1}^{N} V_j \left[G_{ij} \sin(\theta_i - \theta_j) - B_{ij} \cos(\theta_i - \theta_j) \right] \quad i = 1,2 \ldots N, \quad j \neq i \quad (3.20)$$

Similarly, real and reactive power flow from bus i to bus j is

$$P_{ij} = V_i^2 \left(g_{si} + g_{ij} \right) - V_i V_j \left[g_{ij} \cos(\theta_i - \theta_j) + b_{ij} \sin(\theta_i - \theta_j) \right] \quad (3.21)$$

$$Q_{ij} = V_i^2 \left(b_{si} + b_{ij} \right) - V_i V_j \left[g_{si} \sin(\theta_i - \theta_j) - b_{ij} \cos(\theta_i - \theta_j) \right] \quad (3.22)$$

where

V_i, θ_i is the voltage magnitude and phase angle at bus i
$G_{ij} + jB_{ij}$ is the complex ijth element of the bus admittance matrix
$g_{ij} + jb_{ij}$ is the series admittance between i and j buses
$g_{si} + jb_{si}$ is the admittance of a shunt branch at bus i

If line current is measured between bus i and j, it can be written as

$$I_{ij} = \frac{\sqrt{(P_{ij}^2 + Q_{ij}^2)}}{V_i}$$

or if the current in the shunt branch is ignored, it can be written as

$$I_{ij} = \sqrt{(\left(b_{ij}^2\right)(V_i^2 + V_j^2 - 2V_iV_j \cos(\theta_i - \theta_j))} \tag{3.23}$$

3.5 WLS State Estimation Algorithm

The state estimation problem is similar to the power flow problem. However, its solution methods are different from the power flow solution methods in two aspects. First, in the SE problem, certain data is either missing or inaccurate, and second, the solution methods may be less exact so as to have high speed processing.

In the WLS method the necessary condition to minimize the performance index is given in (3.14) as $-H^T(x)R^{-1}\left[z - h(x)\right] = 0$. Here H, the Jacobian matrix, can be written as

$$H = \begin{bmatrix} \frac{\partial P_{inj}}{\partial \theta} & \frac{\partial P_{inj}}{\partial V} \\ \frac{\partial P_{flow}}{\partial \theta} & \frac{\partial P_{flow}}{\partial V} \\ \frac{\partial Q_{inj}}{\partial \theta} & \frac{\partial Q_{inj}}{\partial V} \\ \frac{\partial Q_{flow}}{\partial \theta} & \frac{\partial Q_{flow}}{\partial V} \\ \frac{\partial I_{mag}}{\partial \theta} & \frac{\partial I_{mag}}{\partial V} \\ 0 & \frac{\partial P_{inj}}{\partial V} \end{bmatrix} \tag{3.24}$$

The matrix is the diagonal covariance matrix of the measurements given by (3.12). As can be seen from this equation, the weights are defined by the inverse of measurement variances. Since high-quality measurements have smaller variances, they will have higher weights.

The following example is taken from [8] to illustrate the construction of the H matrix.

Example 3.1

The three-bus power system shown in Figure 3.3 is used to illustrate the construction of the H matrix.

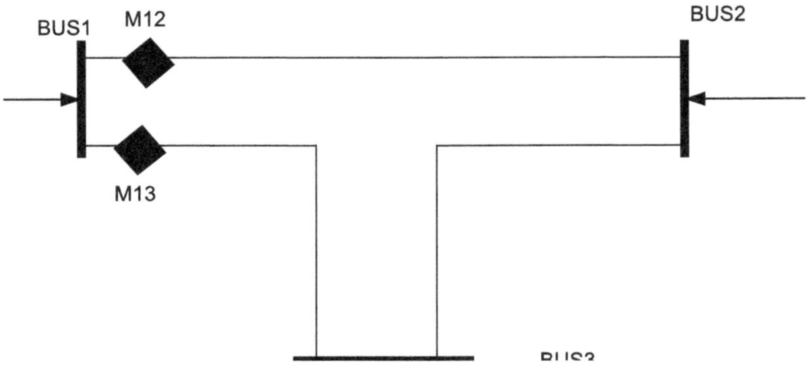

Figure 3.3 Single-line diagram of a three-bus system.

Bus 1 is assumed to be a slack bus, the network data is listed in Table 3.1, and the available measurements are listed in Table 3.2.

There are eight measurements in the system, hence $m = 8$, and the state vector will have five elements or $n = 5$, and $x^T = [\theta_1, \theta_2, V_1, V_2, V_3]$.

All the elements of the H matrix are partial derivatives of available measurements with respect to θ and V. The following conditions will apply:

$$\partial V_i/\partial \theta_j = 0,\ \partial V_i/\partial V_j = 0,\ \text{and}\ \partial V_i/\partial V_i = 1 \tag{3.25}$$

The transmission line is represented by its equivalent network shown in Figure 3.2.

The elements of the H matrix corresponding to real power injections are

$$\partial P_i/\partial \theta_i = \sum_{j=1}^{N} V_i V_j \left(-G_{ij}\sin(\theta_i - \theta_j)\right) + B_{ij}\cos(\theta_i - \theta_j) - V_i^2 B_{ii}$$

$$\partial P_i/\partial \theta_j = V_i V_j \left(G_{ij}\sin(\theta_i - \theta_j)\right) - B_{ij}\cos(\theta_i - \theta_j)$$

Table 3.1
Network Data

From Bus	To Bus	*R* (pu)	*X* (pu)	Susceptance (pu)
1	2	0.01	0.03	0
1	3	0.02	0.05	0
2	3	0.03	0.08	0

Table 3.2
Measurement Values

Measurements	Value (pu)	{T3-2_EQ1}
V_1	1.06	0.004
V_2	0.968	0.004
$p_{1\text{-}2}$	0.888	0.008
$p_{1\text{-}3}$	1.173	0.008
$q_{1\text{-}2}$	0.568	0.008
$q_{1\text{-}3}$	0.663	0.008
P_2	–0.501	0.010
Q_2	–0.286	0.010

$$\begin{aligned} \partial P_i/\partial V_i &= \sum_{j=1}^{N} V_j\left(G_{ij}\cos(\theta_i-\theta_j)\right)+B_{ij}\sin(\theta_i-\theta_j)+V_iG_{ii} \\ \partial P_i/\partial V_j &= V_i\left(G_{ij}\cos(\theta_i-\theta_j)\right)+B_{ij}\sin(\theta_i-\theta_j) \end{aligned} \tag{3.26}$$

The elements for the reactive power injection are

$$\begin{aligned} \partial Q_i/\partial\theta_i &= \sum_{j=1}^{N} V_iV_j\left(G_{ij}\cos(\theta_i-\theta_j)\right)+B_{ij}\sin(\theta_i-\theta_j)-V_i^2G_{ii} \\ \partial Q_i/\partial\theta_j &= V_iV_j\left(-G_{ij}\cos(\theta_i-\theta_j)\right)-B_{ij}\sin(\theta_i-\theta_j) \\ \partial Q_i/\partial V_i &= \sum_{j=1}^{N} V_j\left(G_{ij}\sin(\theta_i-\theta_j)\right)-B_{ij}\cos(\theta_i-\theta_j)-V_iB_{ii} \\ \partial Q_i/\partial V_J &= V_i\left(G_{ij}\sin(\theta_i-\theta_j)\right)-B_{ij}\cos(\theta_i-\theta_j) \end{aligned} \tag{3.27}$$

The real and reactive power flows in line *i-j* depend only θ_i and V_i and can be written as

$$\begin{aligned}
\partial P_{ij}/\partial\theta_i &= V_iV_j\left(g_{ij}\sin(\theta_i-\theta_j)-b_{ij}\cos(\theta_i-\theta_j)\right)\\
\partial P_{ij}/\partial\theta_j &= -V_iV_j\left(g_{ij}\sin(\theta_i-\theta_j)-b_{ij}\cos(\theta_i-\theta_j)\right)\\
\partial P_{ij}/\partial V_i &= -V_j\left(g_{ij}\cos(\theta_i-\theta_j)+b_{ij}\cos(\theta_i-\theta_j)\right)+2(g_{ij}+g_{si})V_i\\
\partial P_{ij}/\partial V_j &= -V_i\left(g_{ij}\cos(\theta_i-\theta_j)+b_{ij}\sin(\theta_i-\theta_j)\right)\\
\partial Q_{ij}/\partial\theta_i &= -V_iV_j\left(g_{ij}\cos(\theta_i-\theta_j)+b_{ij}\sin(\theta_i-\theta_j)\right)\\
\partial Q_{ij}/\partial\theta_j &= V_iV_j\left(g_{ij}\cos(\theta_i-\theta_j)+b_{ij}\sin(\theta_i-\theta_j)\right)\\
\partial Q_{ij}/\partial V_i &= -V_j\left(g_{ij}\sin(\theta_i-\theta_j)-b_{ij}\cos(\theta_i-\theta_j)\right)-2V_i(b_{ij}+b_{si})\\
\partial Q_{ij}/\partial V_j &= -V_i\left(g_{ij}\sin(\theta_i-\theta_j)-b_{ij}\cos(\theta_i-\theta_j)\right)
\end{aligned} \tag{3.28}$$

Ignoring the effect of shunt branch currents, the current magnitude measurements will have the following elements

$$\begin{aligned}
\partial I_{ij}/\partial\theta_i &= \left(\frac{g_{ij}^2+b_{ij}^2}{I_{ij}}\right)V_iV_j\sin(\theta_i-\theta_j)\\
\partial I_{ij}/\partial\theta_j &= -\left(\frac{g_{ij}^2+b_{ij}^2}{I_{ij}}\right)V_iV_j\sin(\theta_i-\theta_j)\\
\partial I_{ij}/\partial V_i &= \left(\frac{g_{ij}^2+b_{ij}^2}{I_{ij}}\right)(V_i-V_j\cos(\theta_i-\theta_j))\\
\partial I_{ij}/V_j &= \left(\frac{g_{ij}^2+b_{ij}^2}{I_{ij}}\right)(V_j-V_i\cos(\theta_i-\theta_j)
\end{aligned} \tag{3.29}$$

To construct the H matrix initially all the elements are put as zeros. Network elements are considered one by one. When line *i-j* is processed it is not necessary to compute all the 16 partial derivatives. The partial derivatives $\partial P_{ij}/\partial\delta_i, \ldots, \partial P_{ij}/\partial\delta_j, \ldots, \partial P_{ij}/\partial V_i$ and $\partial P_{ij}/\partial V_j$ are to be computed only when P_{ij} or P_i or both are in the measurement list.

Solution for Example 3.1

The initial state vectors are defined as the angles at bus 2 and bus 3 and the voltage magnitudes at buses 1, 2, and 3. Taking bus 1 as reference bus and assuming a flat start:

$$[x]^T = [x_1\ x_2\ x_3 x_4 x_5]$$
$$[x^0]^T = \begin{bmatrix}0 & 0 & 1.0 & 1.0 & 1.0\end{bmatrix}$$

The elements of the H matrix will be evaluated using (3.24) through (3.26) as

$$g_{12} = 10,\ b_{12} = -30, g_{13} = 6.89,\ b_{13} = -17.24$$
$$g_{23} = 4.109,\ b_{23} = -j10.95,$$

$$Y_{bus} = \begin{bmatrix} 16.89 - j47.24 & 10 - j30 & 6.89 - j17.24 \\ 10 - j30 & 14.109 - j40.95 & 4.109 - j10.95 \\ 6.89 - 17.24 & 4.109 - j10.95 & 10.999 - 28.29 \end{bmatrix}$$

$$\frac{\partial P_{12}}{\partial \theta_2} = V_1 V_2 \left(g_{12}\sin(\theta_1 - \theta_2) - b_{12}\cos(\theta_1 - \theta_2)\right) = -b_{12} = -30.0$$
$$\partial P_{12}/\partial V_1 = -V_2\left(g_{12}\cos(\theta_1 - \theta_2) + b_{12}\sin(\theta_1 - \theta_2) + 2(g_{12} + g_{s1})V_1\right.$$
$$= -V_2 g_{12} + 2 g_{12} V_1 \ = 10.0$$
$$\partial Q_{12}/\partial \theta_1 = -V_1 V_2\left(g_{12}\cos(\theta_1 - \theta_2) + b_{12}\sin(\theta_1 - \theta_2)\right) = 10.0$$
$$\partial Q_{12}/\partial V_1 = -V_2\left(g_{12}\sin(\theta_1 - \theta_2) - b_{12}\cos(\theta_1 - \theta_2)\right) - 2V_1(b_{12} + b_{s1})$$
$$= 30.0 - 60.0 = 30.0$$
$$\partial P_1/\partial \theta_1 = \sum_{j=1}^{N} V_i V_j \left(-G_{ij}\sin(\theta_i - \theta_j)\right) + B_{ij}\cos(\theta_i - \theta_j) - V_i^2 B_{ii}$$

For calculating the terms related to bus injection, the Y_{bus} matrix is to be formed.

$$Y_{bus} = \begin{bmatrix} Y_{11} & Y_{12} & Y_{13} \\ Y_{21} & Y_{22} & Y_{23} \\ Y_{31} & Y_{32} & Y_{33} \end{bmatrix}$$

where

$$Y_{11} = y_{10} + y_{12} + y_{13} = 0 + 10 - j30 + 6.89 - j17.24$$
$$16.89 - j47.24$$
$$Y_{12} = Y_{21} = -y_{21} = -10 + j30$$
$$Y_{bus} = \begin{bmatrix} 16.89 - j47.24 & -10 + j30 & -6.89 + j17.24 \\ -10 + j30 & 11 - j28.2 & -4.109 + j10.95 \\ -6.89 + j17.24 & -4.109 + j10.95 & 14.109 - j40.95 \end{bmatrix}$$
$$\partial P_2/\partial \theta_2 = V_1V_2\left(-G_{12}\sin(\theta_1 - \theta_2\right) + B_{12}\cos(\theta_1 - \theta_2) - V_1^2 B_{11}$$
$$V_1V_3\left(-G_{13}\sin(\theta_1 - \theta_3\right) + B_{13}\cos(\theta_1 - \theta_3) - V_1^2 B_{11}$$
$$= 40.9$$

Similarly other terms may be calculated. The H matrix will be

$$[H] = \begin{bmatrix} -30 & 0 & 10 & -10 & 0 \\ 0 & -17.2 & 6.9 & 0 & -6.9 \\ 40.9 & -10.9 & -10.0 & 14.1 & -4.1 \\ 10 & 0 & 30 & -30 & 0 \\ 0 & 6.9 & 17.24 & 0 & -17.24 \\ -14.1 & 4.11 & -30 & 40.96 & -10.96 \\ 0 & 0 & 1 & 0 & 0 \\ 0 & 0 & 0 & 1 & 0 \end{bmatrix}$$

The gain matrix $G = H^T R^{-1} H$.

$$[R]=\begin{bmatrix} R_{11} & & & & & & & \\ & R_{22} & & & & & & \\ & & R_{33} & & & & & \\ & & & R_{44} & & & & \\ & & & & R_{55} & & & \\ & & & & & R_{66} & & \\ & & & & & & R_{77} & \\ & & & & & & & R_{88} \end{bmatrix}$$

$$R=\begin{bmatrix} 15625 & & & & & & & & \\ & 15625 & & & & & & & \\ & & 10000 & & & & & & \\ & & & 15625 & & & & & \\ & & & & 15625 & & & & \\ & & & & & 10000 & & & \\ & & & & & & 62500 & & \\ & & & & & & & 62500 \end{bmatrix}$$

or

$$G=10^7\begin{bmatrix} 3.439 & -0.506 & 0.0137 & 0 & -0.9137 \\ -0.506 & -0.675 & -0.0137 & 0.0137 & 0 \\ 0.0137 & -0.0137 & 3.107 & -2.9324 & -0.168 \\ 0 & 0.0137 & -2.93 & 3.445 & -0.506 \\ -0.0137 & 0 & -0.168 & 0.506 & 0.675 \end{bmatrix}$$

Using zero as an initial guess for voltage angles and the measured voltages, the initial state vector is voltage at bus 3 taken as 1.0.

$$\begin{bmatrix} x_1^{(0)} \\ x_2^{(0)} \\ x_3^{(0)} \\ x_4^{(0)} \\ x_5^{(0)} \end{bmatrix}=\begin{bmatrix} 0 \\ 0 \\ 1.006 \\ 0.968 \\ 1.0 \end{bmatrix}$$

Now $z = h(x)$. There are eight measurements that can be written in terms of states as

$$V_1 = x_3, V_2 = x_4, P_{12} = V_1^2\left(g_{s1} + g_{12}\right) - V_1^2\left(g_{s1} + g_{12}\right) -$$
$$V_1V_2\left[g_{12}\cos(\theta_1 - \theta_2) + b_{12}\sin(\theta_1 - \theta_2)\right]$$
$$P_{12} = 10x_3^2 - 10x_3x_4\cos x_1 - 30\sin x_1 = 0$$
$$Q_{12} = -V_1^2\left(b_{s1} + b_{12}\right) - V_1V_2\left[g_{12}\sin(\theta_1 - \theta_2) + b_{12}\cos(\theta_1 - \theta_2)\right]$$
$$= 30x_3^2 + 10x_3x_4\sin x_1 - 30\cos x_1 = 0$$

and

$$P_2 = 10x_3x_4\cos x_1 - 30x_3x_4\ \sin x_1 + 4.109\,x_4x_5\cos(x_1 - x_2) -$$
$$10.95x_4x_5\ \sin(x_1 - x_2)$$

and

$$[z] = \begin{bmatrix} 1.006 \\ 0.968 \\ 0.888 \\ 1.173 \\ 0.568 \\ 0.663 \\ -0.501 \\ -0.286 \end{bmatrix} = \begin{bmatrix} x_3 \\ x_4 \\ 10x_3^2 - 10x_3x_4\cos x_1 - 30\sin x_1 \\ 6.89x_3^2 - 6.89x_3x_5\cos x_2 - 17.24\sin x_2 \\ 30x_3^2 - 10x_3x_4\sin x_1 - 30\cos x_1 \\ 6.89x_3^2 - 6.89x_3x_5\sin x_2 - 17.24\cos x_2 \\ 10x_3x_4\cos x_1 - 30x_3x_4\, sinx_1 + 4109\,x_4x_5\cos(x_1 - x_2) - \\ 10.95\ x_4x_5\, sin(x_1 - x_2) \end{bmatrix}$$

Using (3.18)

$$x^1 = x^0 + \left(H(x^0)^T R^{-1} H(x)^0\right)^{-1} \quad H\left(x^0\right)^T R^{-1}\left[z - h\left(x^0\right)\right]$$

After the first iteration

$$\begin{bmatrix} x_1 \\ x_2 \\ x_3 \\ x_4 \\ x_5 \end{bmatrix} = \begin{bmatrix} -0.02 \\ -0.045 \\ 0.9997 \\ .9743 \\ 0.9428 \end{bmatrix}$$

Finally after three iterations, the values obtained are

$$\begin{bmatrix} -1.2475 \\ -2.7475 \\ 0.9996 \\ 0.9741 \\ 0.9439 \end{bmatrix}$$

The same problem can be solved using Cholesky's method. The gain matrix can be decomposed into a lower triangular sparse matrix and its transpose. Using this transformation, the solution to state equations can be obtained by forward and backward substitution.

$$L L^T \Delta x_k = H\left(x^k\right)^T R^{-1}\left[z - h\left(x^k\right)\right] \quad (3.30)$$

Forward Substitution

The elements of matrix $[L^T \Delta x^k]$ = [u] are obtained starting with first row and equating it to the first row of $H(x^k)^T R^{-1}[z - h(x^k)]$ will yield a solution. Substituting the value of Δ will result in reduction of variables by 1. Similarly, solving for other rows, the solution of the matrix $[u]$ will be obtained.

Back Substitution

After the matrix $[u]$ is known, the value of Δx^k is obtained by back substitution starting with the last row. Thus by using forward and backward substitution the Δx^k are obtained until the difference is within the small value specified.

3.5.1 State Estimation by Orthogonal Decomposition

In the WLS method of state estimation, the solution to the system states using normal equation may sometimes have very slow convergence or may fail to converge under certain conditions. Analysis has shown that ill-conditioning may happen when any of the following factors are present:

- Disparity in weighting factors;
- Large number of injection measurements;
- Connection of long and short transmission lines.

The orthogonal transformation method (also known as the Q-R method) performs Q-R decomposition of the Jacobian matrix directly. According to this, any *m-n* matrix of full rank can be decomposed into two matrices of the form

$$\mathbf{A} = \mathbf{Q}\,\mathbf{R}$$

Where **Q** is an orthogonal matrix of size *m-m* and **R** is an upper triangular matrix of size *m-n*. Since **Q** is an orthogonal matrix $Q^T Q = I$, or

$$Q^T A = R \tag{3.31}$$

If Q and R are partitioned as

$$A = \begin{bmatrix} Q_n & Q_0 \end{bmatrix} \begin{bmatrix} U \\ 0 \end{bmatrix} = Q_n U \tag{3.32}$$

the orthogonal transformation is therefore more numerically stable. When this algorithm is applied to the WLS state estimation, (3. 17) $\Delta x = (H^T R^{-1} H)^{-1} H^T R^{-1}[z - h(x)]$ is solved to eliminate the R^{-1} matrix.

If the R^{-1} matrix is written as

$$R^{-1} = R^{-\frac{1}{2}} R^{-\frac{1}{2}} \tag{3.33}$$

where

$$R^{-\frac{1}{2}} = \begin{bmatrix} \frac{1}{\sigma_{m1}} & & \\ & \frac{1}{\sigma_{m2}} & \\ & & \frac{1}{\sigma_{mn}} \end{bmatrix} \tag{3.34}$$

or

$$\left[H^T R^{-1} H\right]^{-1} = H^T R^{-\frac{1}{2}} R^{-\frac{1}{2}} H]^{-1} = \left[H'^T \quad H'\right] \tag{3.35}$$

where

$$[H'] = [R^{-1/2}][H]$$

Equation (3.17) therefore becomes

$$\Delta x = (H'^T H')^{-1} (H'^T) \left[R^{-1/2} \right] [z - h(x)] \tag{3.36}$$

or

$$\Delta x = (H'^T H')^{-1} (H'^T) [\Delta z'] \tag{3.37}$$

where

$$\Delta z' = \left[R^{-1/2} \right] [\Delta z]$$

and

$$\Delta z = [z - h(x)]$$

The orthogonal decomposition algorithm is used to find an orthogonal matrix $[Q]$ such that

$$[H'] = [Q][U] \tag{3.38}$$

The matrix $[U]$ has an upper triangular structure. The state estimation equation with these substitutions become

$$\Delta x = [U^T Q^T Q U]^{-1} [U^T Q^T] \Delta z' \tag{3.39}$$

which after rearrangement can be written as

$$[U^T][U] \Delta x = [U^T][Q^T] \Delta z' \tag{3.40}$$

or

$$[U] \Delta x = [Q^T] \Delta z' \tag{3.41}$$

The orthogonal decomposition algorithm for the power system state estimation can be explained as

- Determine the Q matrix and U matrix. The Q matrix must be orthogonal.
- Perform the factorization $[H'] = [Q][U]$.
- Compute the vector $[Q^T \Delta z']$.
- Compute x from back substitution using $[U]\Delta x = [Q^T]\Delta z'$.

3.5.2 Equality Constrained State Estimation

Most power systems contain zero injection buses at which there is no load or generation. These buses impose constraints on a state estimator, since the real and reactive power injected at these buses is fixed at zero. The constrained WLS state estimation problem can be formulated as

$$\min J(x)\left[z - h(x)^T\right][R^{-1}][z - h(x)] \tag{3.42}$$

with equality constraints $(x) = 0$. These constraints represent the zero injection measurements (virtual measurements) that are excluded from $h(x)$.

This problem is treated as an optimization problem and a Lagrangian function may be constructed as

$$\mathcal{L}(x, \lambda) = J(x) - \lambda^T c(x) \tag{3.43}$$

The optimal solution of the above equation can be obtained by taking the first-order differentiation and equating it to zero (Kuhn-Tucker method).

$$\frac{\partial \mathcal{L}}{\partial x} = 0 = -H^T(x)R^{-1}[z - h(x)] - \boldsymbol{C}^T\lambda = 0 \tag{3.44}$$

where $\boldsymbol{C} = \partial c / \partial x$ and $c(x) = 0$.

Using Taylor series expansion the above equations can be linearized at the kth iteration as

$$\begin{bmatrix} H'(x^k)H(x^k) & -\boldsymbol{C}'(x^k) \\ -\boldsymbol{C}(x^k) & 0 \end{bmatrix} \begin{bmatrix} \Delta x^k \\ \lambda^{k+1} \end{bmatrix} = \begin{bmatrix} H'(x^k)\Delta z^k \\ -cx(k) \end{bmatrix} \tag{3.45}$$

The equality constraints provide two disadvantages: (1) the coefficient matrix in the above equation is indefinite, and (2) if the constraint has bad data, the convergence is difficult to obtain.

The advantages of constrained optimization is that the C matrix is not squared (i.e., C^TC is not required). As well, no weights are assigned to the equality constraints so these can be forced without affecting the coefficient matrix.

3.6 Decoupled State Estimation Method

In the WLS estimation method described earlier, assignment of large weights and round-off errors cause numerical problems and may sometimes make the system ill-conditioned. The calculation and triangular decomposition of the matrix also has a very large computation time. Efficient fast decoupled state estimation techniques based on P-δ and Q-V decoupling have been developed based on fast decoupled load flow techniques. For large-scale power systems the transmission lines have a very high X/R ratio. In such a case the real power changes are less sensitive to change in voltage magnitude and can be ignored. Similarly the reactive power change is less sensitive to changes in angles. Making these simplifications, the Gain matrix in state estimation algorithm can be simplified.

The measurement matrix is therefore partitioned in real power measurements and reactive power measurements. The real power measurements include the real power bus injections and real power flows in branches. The reactive power measurements include reactive power bus injections, reactive power flows in branches, and bus voltage magnitude. The two main characteristics of fast decoupled state estimators are

1. Active/reactive decoupling;
2. Use of constant matrices (G and/H, which are functions of network admittance only).

The active/reactive decoupling can be applied only on the G matrix (*algorithm decoupling*) or also on the Jacobian matrix H (*model decoupling*).

An algorithm decoupled state estimator use the same approximations in gain matrix as a fast decoupled load flow. The gain matrix can be partitioned as

$$G(V,\theta)=\begin{bmatrix} G_{11} & G_{12} \\ G_{21} & G_{22} \end{bmatrix} \quad (3.46)$$

In the case of decoupling, G_{12} and G_{21} are zero. Two different approaches can be made in calculating G_{11} and G_{22}. Starting with $V= 1.0$ p.u. and θ or $V= 1.0$ p.u. and θ and resistance =0.

The Jacobian matrix H is partitioned as

$$H(V,\theta)=\begin{bmatrix} H_{11} & H_{12} \\ H_{21} & H_{22} \end{bmatrix} \tag{3.47}$$

3.6.1 Algorithm Decoupling

In algorithm decoupling, G_{21} and G_{21} only are assumed to be 0. The terms and are calculated using the H matrix. The components of the H matrix are calculated with a flat start. The new gain matrix is a function of network admittances only.

$$G=\begin{bmatrix} G' & 0 \\ 0 & G'' \end{bmatrix} \tag{3.48}$$

3.6.2 Model Decoupling

In model decoupling, approximations are introduced to decouple the Jacobian matrix and to make it constant within network admittances only. The advantage of this method is that H is not calculated at each iteration.

3.7 DC State Estimator

State estimation using ac power flow model considers both real and reactive power and is formulated by nonlinear equations. The dc state estimation problem is based on an approximate mathematical model similar to the model used in dc power flow. In dc state estimation many simplifications are made. First, it is assumed that all the bus voltages have a magnitude of 1.0 per unit. The line is assumed to be lossless and the angle between the voltages at the buses is very small. As well, all the shunt elements are neglected. Under these approximations the power flowing on each line using dc power flow is

$$P_{ij} = \frac{1}{b_{ij}}\left(\delta_i - \delta_j\right) + e \qquad (3.49)$$

where b_{ij} is the reactance of $i - j$ branch, and e is the error in measurement.

Similarly the power injection at bus is given by

$$P_i = \sum_{j=1}^{n} P_j + e \qquad (3.50)$$

where is the number of buses connected to bus j.

The dc model for the real power measurements can be expressed in terms of states as

$$z_{meas} = \boldsymbol{H}\delta + e \qquad (3.51)$$

where $\boldsymbol{H}$ is a function of branch reactance only, and $\boldsymbol{e}$ is a vector of random errors.

Example 3.2

Determine states of the three-bus power system shown in Figure 3.3 with the following measurement. Use dc state estimation to determine the angles at bus 1 and 2.

Line data; $x_{12} = 0.25$, $x_{13} = 0.4$, $x_{23} = 0.2$.

The measured values z are M12 = 52Mw, M13 = 16Mw and M23 = 28 Mw.

Solution

First we find the H matrix. For that we write measurements as functions of state variables. Assuming 100 Mw as base value for power.

Using (3.49):

$$M_{12} = 0.52 = \frac{1}{0.25}(\delta_1 - \delta_2) = 4\delta_1 - 4\delta_2$$

$$M_{13} = 0.16 = \frac{1}{0.40}(\delta_1 - \delta_3) = 2.5\delta_1$$

$$M_{32} = 0.28 = \frac{1}{0.20}(\delta_3 - \delta_2) = -5\delta_2$$

$$H = \begin{bmatrix} 4 & -4 \\ 2.5 & 0 \\ 0 & -5 \end{bmatrix}$$

assuming the errors in 3 meters is the same $\sigma = 0.01$ or $\frac{1}{\sigma^2} = 0.0001.$

$$\text{The covariance matrix } R = \begin{bmatrix} 0.0001 & & \\ & 0.0001 & \\ & & 0.0001 \end{bmatrix}$$

The estimated values of angles are using the WLS state estimator are given by

$$\hat{x} = \left[(H)^T (R)^{-1} (H)\right]^{-1} [H]^T [R^{-1}]\, z$$

or

$$\begin{bmatrix} \delta_1 \\ \delta_2 \end{bmatrix} = \begin{bmatrix} 4 & 2.5 & 0 \\ -4 & 0 & -5 \end{bmatrix} \begin{bmatrix} 0.0001 & & \\ & 0.0001 & \\ & & 0.0001 \end{bmatrix}^{-1} \begin{bmatrix} 4 & -4 \\ 2.5 & 0 \\ 0 & -5 \end{bmatrix}]^{-1} \times$$

$$\begin{bmatrix} 4 & 2.5 & 0 \\ -4 & 0 & -5 \end{bmatrix} \begin{bmatrix} 0.0001 & & \\ & 0.0001 & \\ & & 0.0001 \end{bmatrix}^{-1} \begin{bmatrix} 0.52 \\ 0.16 \\ 0.28 \end{bmatrix}$$

The measured values z are obtained from meters with errors.

We get

$$\begin{bmatrix} 0.00792 \\ -0.0548 \end{bmatrix}$$

From estimated angles the estimated power can be calculated.

References

[1] Schweppe, F. C., and J. Wildes, "Power System Static State Estimation, Part1—Exact Model," *IEEE Transactions on Power Apparatus and Systems,* Vol. PAS-89, 1970, pp. 120–125.

[2] Schweppe, F. C., J. Douglas, and D. Rom, "Power System Static State Estimation, Part II—Approximate Model," *IEEE Transactions on Power Apparatus and Systems,* Vol. PAS-89, 1970, pp. 125–130.

[3] Monticelli, A., "Electric Power System State Estimation," *Proceedings of the IEEE,* Vol. 88, No. 2, February 2000, pp. 262–282.

[4] Wu, F. F., "Power System State Estimation: A Survey," *International Journal of Electric Power and Energy Systems,* Vol. 12, No. 2, April 1990, pp. 80–87.

[5] Larson, R. E., W. F. Tinney, and J. Peschon, "State Estimation in Power System, Part I—Theory and Reliability," *IEEE Transactions on Power and Apparatus Systems,* Vol. 89, 1970, pp. 345–352.

[6] Schweppe, F. C., and E. J. Handschin, Static State Estimation in Electric Power System," *Proceedings of IEEE,* Vol. 62, 1974, pp. 972–982.

[7] Monticelli, A., *State Estimation in Electric Power Systems: A Generalized Approach,* Norwell, MA: Kluwer, 1999.

[8] Abur, A., and A. G. Exposito, *Power System State Estimation: Theory and Implementation* New York: Marcel Dekker, 2004.

[9] Wood, A. J., and B. F. Wollenberg, *Power Generation Operation and Control,* New York: John Wiley and Sons, 1985.

4

Network Observability and Pseudomeasurements

4.1 Network Graphs and Matrices

The objective of a state estimation is to obtain a computer model that accurately represents the current conditions in a power system. It is a data processing scheme that computes the state of a system from the following three pieces of information:

1. Measurement of system variables;
2. Mathematical model of the system (includes the system topology);
3. Prior knowledge of system inputs and outputs known as pseudomeasurements.

When set of available measurements are sufficient to calculate the state vector of the system, it is termed as an observable system [1–4]. For real-time monitoring of a power system, an observability test should be carried out prior to state estimation. If the system is observable the state estimation may be carried out straight away. However, in an unobservable system, there may be observable islands and unobservable regions within the network. The states in unobservable regions of the network are estimated by adding additional measurements using pseudomeasurements. Pseudomeasurements are not obtained from meters but are typically calculated using historical data or short-term load forecast. The most common exact pseudomeasurement is the bus injection at a substation that has no generation and serves no load. Here we can say with certainty that real and reactive power injection at this bus is zero.

Thus, a network observability test should be performed prior to executing the state estimation. Since this problem is related to system configuration or network topology, analysis of topological observability is performed to determine the network observability. The analysis of network observability makes extensive use of graph theory, which has been used in a network represented by lines. In graph theory the elements of a network are represented by lines called *edges* or *branches* and their connecting points called *nodes.* The graph of a physical system consists of set of nodes and edges, where each edge is connected to two different nodes. A linear graph depicts the geometrical interconnection of the elements of a given network. A graph is said to be fully connected if all the nodes are connected by edges and any node can be reached from any other node by tracing the edges of the graph. This graph is known as undirected graph. A directed graph is a graph where all the edges are assigned directions. The direction of the edge is identified by an arrow from the sending end node to the receiving end.

Tree: A *tree* of the graph is defined as connected graph containing all the nodes but no closed path or loops. If n is the number of nodes then the number of branches (edges) in the tree is (n–1). The elements of the original graph not included in the tree form a graph that may not be necessarily connected is known as a *cotree.* All other branches that are not part of the tree are included in *cotree* and are called *links.* For a graph with n nodes and e edges, there will be $(e - n + 1)$ links associated with this tree. In a tree of a graph the number of edges are $(n - 1)$ but the set of edges that form the tree are not unique. That means there can be more than one tree of a graph. If a link is added to the tree, the corresponding graph contains one closed path called a basic loop. Thus, a graph has as many basic loops as the number of links.

Cut-set: Cut-set is the minimum number of elements in the connected graph that if removed will divide the graph in only two connected subgraphs. *Basic cut-sets* are those that contain only one branch. Hence the number of basic cut-sets is equal to the number of branches.

Matrix representation of a graph: Graphs can be represented and effectively stored in the computer using matrices. The following types of matrices are used to represent graphs:

- *Incidence matrix:* If a network is represented by graph G with n nodes and e elements (branches), then the matrix A with n rows and e columns is known as *node to branch* incidence *matrix*. The matrix elements are given by:

$A_{ij} = 1$, if the *j*th element is incident to but directed away from the node *i*

$A_{ij} = -1$ if the *j*th element is incident to and directed towards node *i*

$A_{ij} = 0$, if the *j*th element is not incident to node *i*

- *Reduced node to branch incidence matrix:* Any node of the network graph can be selected as reference node and the matrix obtained by deleting the row corresponding to reference node is known as *reduced node to bus incidence matrix* (A_r) for a power system network. Here the nodes are termed as buses and elements as branches. A loop-free subgraph *F* of *G* is called a forest.

Example 4.1

A single diagram of a power system is shown in Figure 4.1. The graph of the network is shown in Figure 4.2. Here, $n = 4$ and $l = 9$. The incidence matrix **A** is shown in Table 4.1.

Taking node zero as a reference node, the reduced node to branch incidence matrix is shown in Table 4.2.

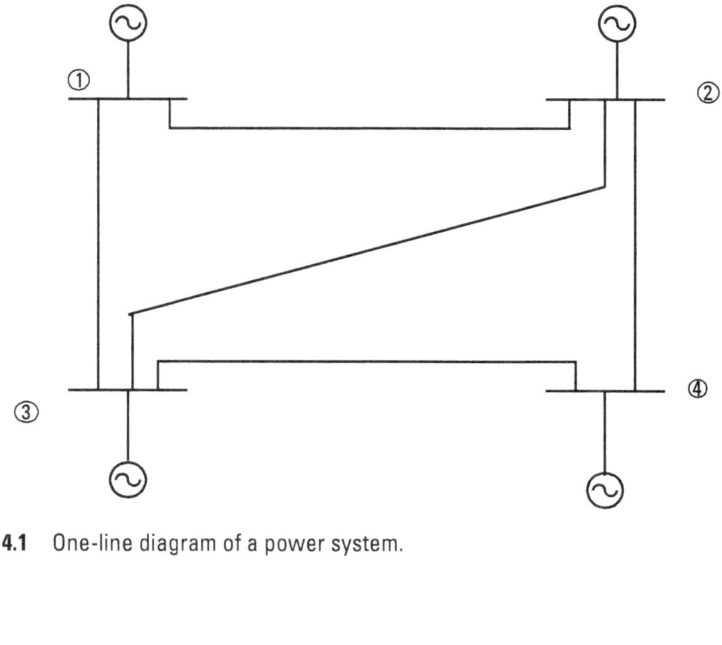

Figure 4.1 One-line diagram of a power system.

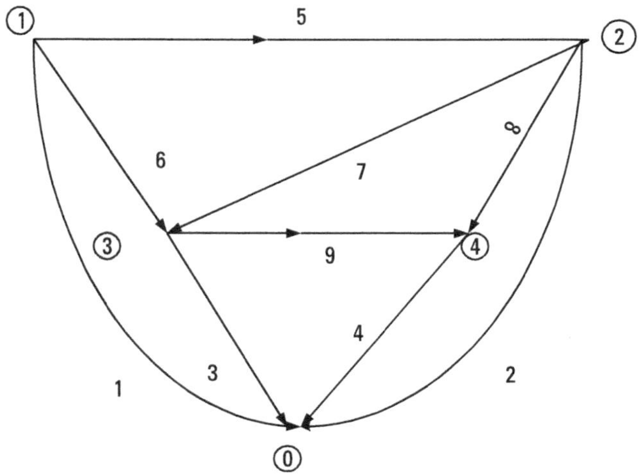

Figure 4.2 Oriented linear graph of the power system shown in Figure 4.1.

Table 4.1
Incidence Matrix A

Elements	*Nodes* 1	2	3	4	0
1	1	0	0	0	–1
2	0	1	0	0	–1
3	0	0	1	0	–1
4	0	0	0	1	–1
5	1	–1	0	0	0
6	1	0	–1	0	0
7	0	1	–1	0	0
8	0	1	0	–1	0
9	0	0	1	–1	0

4.2 Bus Admittance and Bus Impedance Matrices

The data obtained from a power system network is in the form of uncoupled elements is called a *primitive network.* The primitive network provides the information regarding individual elements only and can be in the form of

Table 4.2
Node to Branch Incidence Matrix

Elements	*Nodes* 1	2	3	4
1	1	0	0	0
2	0	1	0	0
3	0	0	1	0
4	0	0	0	1
5	1	–1	0	0
6	1	0	–1	0
7	0	1	–1	0
8	0	1	0	–1
9	0	0	1	–1

$A_r =$ (matrix above)

an admittance or impedance network. The impedance form of a primitive network is shown in Figure 4.3(a).

The performance equation of a primitive network in impedance form is:

$$v_{ij} + e_{ij} = Z_{ij} i_{ij} \tag{4.1}$$

where $v_{ij} = v_i - v_j$ and v_i and v_j are the node voltages.

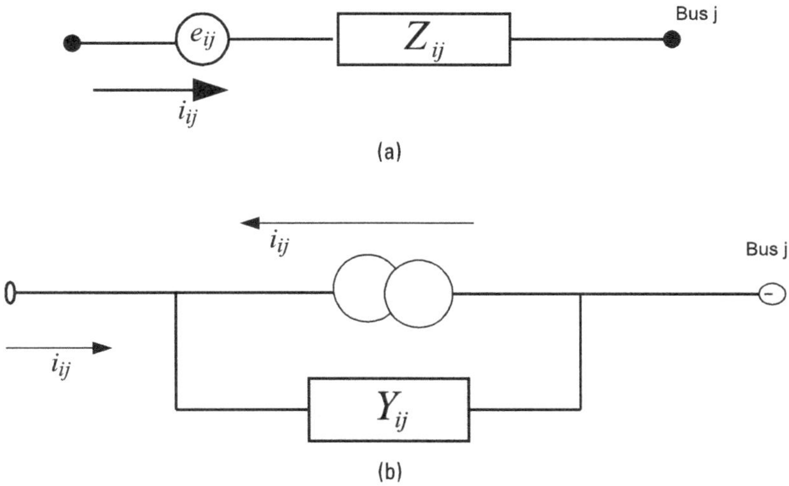

Figure 4.3 Primitive network: (a) impedance and (b) admittance.

Similarly, in the admittance form the primitive network is as shown in Figure 4.3(b). The performance equation can be written as

$$i_{ij} + j_{ij} = Y_{ij} v_{ij} \tag{4.2}$$

The performance equation of the entire network can be written in the form obtained in (4.1) and (4.2) as

$$[v] + [e] = [Z][i] \tag{4.3}$$

and

$$[i] + [j] = [Y]\,[v] \tag{4.4}$$

Here the diagonal elements of matrices Z and Y are the self-impedances and admittances of the elements, respectively. Similarly, the off-diagonal elements of these matrices are mutual impedances and admittances, respectively. If there is no mutual coupling, the matrix Z or Y will be diagonal.

4.2.1 Loop to Branch Incidence Matrix

As discussed in (4.1), a tree of a graph is defined as set of connected edges containing all the nodes but no close path. If any node in the graph can be reached from any other node by tracing only the edges of the tree, then it is known as a spanning tree of the graph. All edges that do not belong to the spanning tree are called *links*. Since each link forms a loop with the tree branches, where no other link is present, each link belongs to the fundamental loop of the graph.

A network with N nodes and L edges will have a spanning tree with $(N - 1)$ edges. There will be $(L - N + 1)$ fundamental loops of the graph (equal to the number of links).

Each fundamental loop includes several branches and a single link. Each loop is assigned a clockwise or counterclockwise direction. The same direction is also applied to the link. With a particular direction assigned, the loop to branch incidence matrix $\mathcal{L} = (L - N + 1) \times L$ can be written as

$$\mathcal{L}(i,j) = \begin{bmatrix} 1 \text{ if branch } j \text{ is in loop } i \text{ and has same direction} \\ -1 \text{ if } j \text{ is in loop } i \text{ with opposite direction} \\ 0 \text{ otherwise} \end{bmatrix}$$

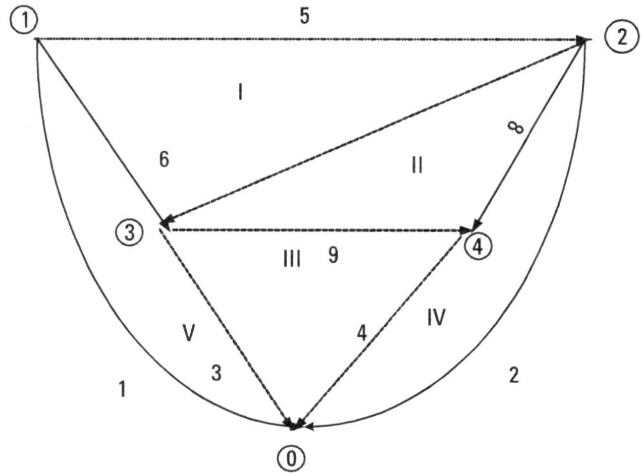

Figure 4.4 Tree of the graph of power system of Figure 4.1.

The matrix can be partitioned as

$$\mathcal{L} = [\mathcal{L}_T \quad I_L] \tag{4.5}$$

where I_L is the identity matrix and the columns of $\mathcal{L}_T$ correspond to spanning tree branches.

Example 4.2

For the same network graph shown in Figure 4.2, the tree of this graph is shown in Figure 4.4. Write the fundamental loop matrix.

Solution

The tree consists of branches 1, 2, 6, and 8. The links are 3, 4, 5, 7, and 9.

There are five loops. Loop I consists of branches 1, 2, and 5; loop II consists of 1, 3, and 6; loop III is 6, 9, 8, 2, and 1; loop IV is 6, 7, 2, and 1; and loop V is 4, 2, and 8. The fundamental loop matrix will have five rows and nine columns as shown below:

$$\mathcal{L}(5,9)=\begin{bmatrix} -1 & 1 & 0 & 0 & 1 & 0 & 0 & 0 & 0 \\ -1 & 0 & 1 & 0 & 0 & 1 & 0 & 0 & 0 \\ -1 & 1 & 0 & 0 & 0 & 1 & 0 & -1 & 1 \\ 1 & -1 & 0 & 0 & 0 & -1 & 1 & 0 & 0 \\ 0 & -1 & 0 & 1 & 0 & 0 & 0 & 1 & 0 \end{bmatrix}$$

4.3 Loop Equations

Loop equations can be used to determine the loop impedance and loop admittance matrices. Given a spanning tree and the corresponding links, a linearly independent set of fundamental loop equations can be written according to Kirchhoff's law as

$$\mathcal{L}\cdot V_b = 0 \tag{4.6}$$

where $\mathcal{L}$ is fundamental loop to branch incidence matrix and V_b is the branch voltage vector. Care should be taken to assign the polarities of the branch voltages, which should be consistent with the assigned branch directions while constructing the $\mathcal{L}$ matrix.

4.4 Observability Analysis

To estimate the system states completely, a minimum amount of real-time data is necessary. An analytical way of determining whether a given set of data is enough for estimating the states completely is called the *observability analysis.* Network observability analysis can be carried out using either numerical or topological approaches [4–7]. Topological approaches use the decoupled measurement model and graph theory. Questions regarding network observability arise in state estimation as the state estimation can only be carried out if the system is observable. In case the system is not observable, pseudomeasurements must be added to the measurement set to allow an estimate of the state vector to be computed. Pseudomeasurements are also necessary to identify the gross errors that may not be eliminated by bad data detection routines.

Topological approaches use the decoupled measurement model and graph theory. Numerical approaches may use fully coupled or decoupled models.

The objectives of topological observability are as follows:

1. To judge whether the network is observable;
2. To identify the observable island if the network is not observable;
3. To identify branches where pseudomeasurements are necessary to make the system observable.

If some of the measurements in a power system are lost or have gross measurement errors, it is important to know whether the system remains observable or not. Observability analysis is performed to determine the critical measurement sets. The methods to identify critical measurement and critical sets are (1) based on graph theory, or (2) numerical methods based on statistical theory. The topological methods are relatively complex and do not allow a direct critical set determination. The numerical methods are conceptually simpler but may present difficulty in numerical solutions as they require the analysis of the measurement residuals.

4.5 Branch Variable Formulation

A network observability problem follows the topological concept of forming a spanning tree of the network. In numerical methods of observability analysis, the measurement and network loop equations are formed numerically and their solution is determined. Numerical approaches may use fully coupled or decoupled model. These methods use the information (gain) matrix or the Jacobian measurement as the model reflecting the configuration of the system and the measurement set. This approach is easy to use due to built-in factorized matrices of the state estimation problem and solution techniques.

4.5.1 New Branch Variables

The real and reactive power flows *i-j* as shown in Figure 4.5 can be expressed as

$$P_{ij} = V_i^2 g_{ij} - V_i V_j \left(g_{ij} \cos\delta_{ij} + b_{ij} \sin\delta_{ij} \right) \tag{4.7}$$

$$Q_{ij} = -V_i^2 b_{ij} - V_i V_j \left(g_{ij} \sin\delta_{ij} - b_{ij} \cos\delta_{ij} \right) \tag{4.8}$$

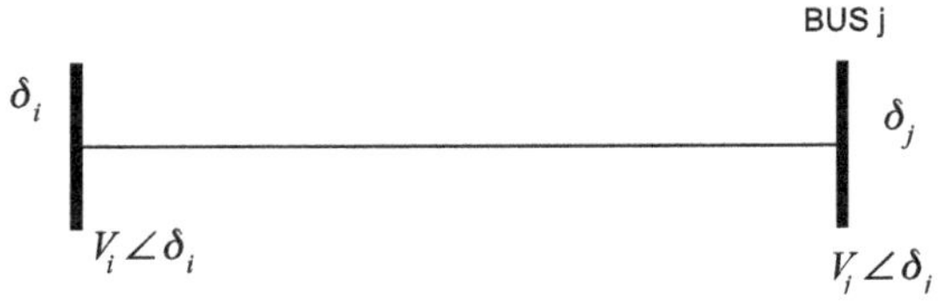

Figure 4.5 A two-bus system with branch variables.

where $g_{ij} + b_{ij}$ is the branch admittance of line *i-j*, neglecting the line-charging susceptance.

V_i and V_j is the magnitude of bus *i* and bus *j*, respectively
δ_{ij} is the branch voltage phase angle = $(\delta_i - \delta_j)$
δ_i and δ_j are the voltage phase angle at bus *i* and bus *j*

Let the branch voltage variables α_{ij} be defined as

$$\alpha_{ij} = \ln\left(\frac{V_i}{V_j}\right) \text{ or } \frac{V_i}{V_j} = e^{\alpha_{ij}} \tag{4.9}$$

The scaled power flow equations can be expressed as

$$P_{ij}^s = g_{ij}\left(1 - e^{\alpha_{ij}} \cos\delta_{ij}\right) - b_{ij}e^{\alpha_{ij}} \sin\delta_{ij} \tag{4.10}$$

$$Q_{ij}^s = -b_{ij}\left(1 - e^{d_{ij}} \cos\delta_{ij}\right) - g_{ij}e^{d_{ij}} \sin\delta_{ij} \tag{4.11}$$

Similarly scaled power injections can be represented as

$$P_i = V_i \sum_{j=1}^{n} V_j \left(g_{ij} \cos\delta_{ij} + b_{ij} \sin\delta_{ij}\right) \tag{4.12}$$

$$P_i^s = \sum_{j=1}^{n} e^{\alpha} \left(g_{ij} \cos\delta_{ij} + b_{ij} \sin\delta_{ij}\right) \tag{4.13}$$

$$Q_i^s = \sum_{j=1}^{n} e^{\alpha} \left(g_{ij} \sin\delta_{ij} - b_{ij} \cos\delta_{ij} \right) \tag{4.14}$$

where n = number of buses directly connected to bus i.

4.5.1.1 Voltage Magnitude

If one of the buses where voltage measurement is available is selected as a reference bus, a network tree that is rooted to this bus can be constructed. The ratio of voltage magnitude at any bus to the voltage magnitude at reference bus can be written as

$$\frac{V_j}{V_0} = V_j^s = \frac{V_j}{V_k} \cdot \frac{V_k}{V_L} \cdots \frac{V_m}{V_0} \tag{4.15}$$

$$= \left[e^{\alpha_{jk}} e^{\alpha_{kl}} \ldots e^{\alpha_{m0}} \right] \tag{4.16}$$

or

$$I_n V_j^s = \left[\alpha_{jk} + \alpha_{kl} + \ldots \alpha_{mo} \right] \tag{4.17}$$

where $k\text{-}l, \ldots m\text{-}o$ are all branches belonging to the tree rooted to bus j.

4.5.1.2 Current Magnitude

In the present representation the line-charging suceptances are ignored. This assumption may not give correct representation for long lines having significant line charging and metered at both ends. However, most of the lines where ampere measurements are made are short lines that have negligible line charging.

Neglecting the line charging, the square of the current magnitudes in line i-j is

$$I_{ij}^2 = \left(V_i^2 + V_j^2 - 2V_i V_j \cos\delta_{ij} \right)\left(g_{ij}^2 + b_{ij}^2 \right) \tag{4.18}$$

Scaling the (4.18), we get:

$$\frac{I_{ij}^2}{V_i^2} = \left(1 + e^{2\alpha_{ij}} - 2e^{\alpha_{ij}} \cos\delta_{ij} \right)\left(g_{ij}^2 + b_{ij}^2 \right) \tag{4.19}$$

4.5.1.3 Loop Equations

The branch variables should also satisfy $(b - N + 1)$ independent loop equations. Here N equals the total number of buses, and b equals the total number of branches. For each loop the equation can be written as

$$\alpha_{ij} + \alpha_{jk} + \alpha_{kl} + \ldots + \alpha_{mi} = 0 \tag{4.20}$$

$$\delta_{ij} + \delta_{jk} + \ldots + \delta_{mi} = 0 \tag{4.21}$$

where *i-j, j-k, k-l,* and so forth are the branches forming the loop. The sparsity of loop equations depends on the way a tree is selected. Also according to Kirchhoff's law, the sum of the branch voltages in a given loop will add up to zero.

$$\mathcal{L} \cdot V_b = 0$$

which is (4.6), where $\mathcal{L} = (l - N + 1) \times L$ fundamental loop to branch incidence matrix, and V_b is the branch voltage vector.

If voltage vector V_b is replaced by branch voltage magnitude difference vector X_α, or branch phase angle difference vector X_δ, (4.6) can be written as

$$\mathcal{L} \cdot X_\alpha = 0,\ X_\alpha = \left[\alpha_i, \alpha_{j,} \alpha_k \ldots \alpha_m\right] \tag{4.22}$$

and

$$\mathcal{L} \cdot X_\delta = 0\ X_\delta = \left[\delta_i, \delta_j, \delta_k \ \ldots \ \delta_m\right] \tag{4.23}$$

4.5.2 Measurement Model Using Branch Variables

The scaled measurement vector z^s can be expressed in terms of the power flows, power injections, and voltage and current magnitudes in the compact form as [8–10]:

$$\boldsymbol{f}(\boldsymbol{X}) = \boldsymbol{z}^s \tag{4.24}$$

where $X^T = [\delta^T, \alpha^T]$ and δ and α are vectors of branch phase angle and branch voltage variables, respectively. z^s is a scaled measurement vector including zeros for the loop equations.

Since the errors in measurement have no effect on the observability of the system the measurements in z are assumed to be error-free. Linearizing of (4.22) using Taylor's first-order approximation results in

$$H\Delta x = z^s - f(x^0) = \Delta z^s \qquad (4.25)$$

where $H = \dfrac{\partial f(X)}{\partial X}$ is $m \times n$ rectangular Jacobian matrix evaluated at some X^0, and $\Delta x = X - X_0$.

In order to ensure that real power measurements are not used to observe voltage, and reactive power measurements are not used to observe phase angle variables, the power flow and injection measurements are assumed to come in pairs. Hence, the only nonzero entries in both phase angle and voltage variable columns will be that of ampere measurements.

Using the decoupling between real power measurements to voltage and reactive power measurements to phase angles, the Jacobian measurement can be built.

The terms of the Jacobian measurement will be evaluated by taking the first derivative of (4.10) to (4.19).

4.5.3 Observability Analysis for Branch Variable Formulation

The necessary condition of observability is given by rank $(H^T) = 2N - 1$ (the number of independent state variables). The rank of (H^T) does not depend on the quality of the measurement set and thus the network observability will not change if we assume that the measurement set is perfect or otherwise. The solution of equation $\Delta\hat{x} = \left(H^T R^{-1} H\right)^{-1} H^T R^{-1} \Delta z$ is thus equivalent to the solution of (4.25):

$$\Delta z = H\,\Delta x$$

Using row pivoting and applying the Peters-Wilkinson decomposition, matrix $\boldsymbol{H}$ can be decomposed into factors as

$$\boldsymbol{H} = \begin{bmatrix} \boldsymbol{L} \\ \boldsymbol{M} \end{bmatrix} [\boldsymbol{U}] \qquad (4.26)$$

where

$H = m \times n$ is the Jacobian matrix

L is the $m \times n$ lower trapezoidal factor

U is the upper triangular matrix

M is a rectangular matrix

m is the total number of measurements and loop equations

n is the total number of branch variables

If the system is fully observable, this factorization can be carried out numerically by row and column pivoting. If the system is not fully observable, zero pivots will be encountered at those columns corresponding to unobservable branches. Wherever a zero pivot is encountered, it will be replaced by a value of 1.0, and the corresponding entry of the right-side vector will be assigned an arbitrary value. The network is observable if only one zero pivot is encountered in the process. If more than one zero pivot is encountered the system is unobservable. The factor L obtained in this manner will contain one or more artificially added pivot entries.

Equation (4.23) can be rewritten as

$$L \cdot U \Delta x = Z^e \tag{4.27}$$

and

$$M U \, \Delta x = \Delta Z^r \tag{4.28}$$

If the system is observable, then (4.25) will result in a null solution. The set of N top measurements (N= branch variables) is sufficient to analyze observability. For an unobservable system the zero pivots that were replaced by 1.0 entry during factorization will now be set to arbitrary but distinct nonzero values. This process is equivalent to adding new measurements to the corresponding branch variable to the existing measurement.

Equation (4.27) can be reordered so that the rows and columns of artificially introduced nonzero pivots become the last. Thus (4.27) can be written as

$$\begin{bmatrix} L_0 & 0 \\ L_e & I_u \end{bmatrix} \begin{bmatrix} U_0 & U_e \\ 0 & I_u \end{bmatrix} \begin{bmatrix} \Delta x_0 \\ \Delta x_u \end{bmatrix} = \begin{bmatrix} 0 \\ \Delta z_u^e \end{bmatrix} \tag{4.29}$$

where U_e is a rectangular matrix, I_u = identity matrix of dimension, N_u, and N_u is the number of zero pivots encountered and replaced by 1.0 during the factorization of H. The solution of (4.29) results as

$$U_0 \Delta x_0 = -U_e - \Delta z_u^e \tag{4.30}$$

and

$$\Delta x_u = \Delta z_u^e \tag{4.31}$$

The nonzero entries in the solution of (4.31) will correspond to unobservable branches. Once the complete list of unobservable branches is obtained, these branches can be eliminated from the system diagram to get the observable island of the system.

In the part of the system which is not observable, additional measurements must be placed to make the system observable.

4.6 Network Topology Processing

Observability analysis can also be carried out using the topological method. In topological studies the decision is based on logical operations only. It therefore only requires information about the connectivity, measurement type, and their locations. The actual parameters of the network are not required in this process. The topological observability handles the relationship between meter placement and system configurations or network topology. The power system configurations often vary during daily system scheduling. Any error in status data results in a misconfiguration of network elements (topological error) and can affect the results of state estimators significantly. Further, any undetected topological error can lead to serious errors when system security is being considered during contingencies. A real-time model of a power network is very important for monitoring a power system to detect erroneous or missing measurements. The real-time model is constructed from a database that contains the description of network in terms of its parameters such as branch impedances and connectivity combined with real-time measurements. The topology processor picks out the status of circuit breakers and switches from the real-time data, and using the connectivity data from the database, determines the present network topology. This topology and all other measurements are used by the state estimator to solve for the bus voltages. Since the availability of real-time measurements may change because of the failure of telemetering equipment and network topology may also change due to faults in some lines, an observability check is required before going for state estimation. Those parts of the network from which measurements are not available are always unobservable. The observability check examines the normally observable portions and identifies those buses that are unobservable. These buses are then made observable by adding pseudomeasurements or taken out of the state estimator. After the state estimator has solved for the observable network, a check is made to detect the presence of bad data. Once the network topology is known the state estimator assumes it to be correct and starts the processing.

The real-time modeling of power network involves the following steps.

1. Network configuration analysis;
2. Observability analysis;
3. State estimation and bad data processing.

4.7 Network Configuration

There are several methods to convert bus-section circuit breaker topology into bus-branch topology. However, the most commonly used method is by using a tree search algorithm. It consists of three steps executed in sequence as described here. In the first step, the bus section at each substation's voltage levels are processed to determine if they are connected together by closed circuit breakers. At the beginning of the step, each bus section is considered a potential separate bus. At the end of the tree search process, all bus sections connected by closed circuit breakers become part of one bus. Thus each separate bus in the topology is uniquely identified along with the bus section. When initializing the program, every bus section and circuit breaker status is processed to determine the initial topology. In the tracking mode, it is necessary to process only those substations in which circuit breakers changed status since the last execution. If there is no change in the status of the circuit breakers, the topology processing need not be executed. In this case, the subsequent network solution is very fast because the ordered and factorized matrices from the previous cycle may be used again and the starting voltage vector is the solution from the previous cycle.

When there is a change in circuit breaker status, the topology processor produces a new branch/bus-oriented model that is used by the network solution. In this case the matrices are to be reordered and refactorized, and start again with a flat voltage vector profile. It will take much longer compared to the case when there is no status change. However, a status change does not necessarily result in a change in the bus/branch model of the network. If the status change is due to change in generation or load at a bus and there is no change in bus connectivity, the topology processing is not necessary. In cases where status changes cause limited modifications to the bus/branch network, special techniques are available to handle the modifications. Earlier topology processors or conventional processors do not keep track of the changes in bus/branch model from one cycle to next. However, now a topology processor is available that keeps track of network modifications over time.

In the next step, all the energized network islands are identified. Starting from a generator bus, the buses are combined by branches into islands; as buses are added to the island, these are sequentially numbered. Also the number of connections to a bus is counted for use in optimal ordering later. When no more buses are to be added to the first island, a search is made for an unprocessed generator bus. If one such bus is found, it is used to start the search for the next island. If no more generator buses are available, the search is complete and all energized islands are identified. Optimal ordering of the buses can be done at this stage.

In the final step, all the equipment connected to the buses is tabulated. The switching devices are not the only aspects that must be considered; the connection of measurement devices must also be considered. Network connectivity analysis identifies the energized, de-energized, and grounded electrical islands. The network topology processor (NTP) creates two data structures: a bus to branch data structure and a network to device data structure.

In case the networks containing only line flow measurements in which real and reactive measurements occur in pairs, the topological condition for observability is that there exists at least one bus voltage magnitude measurement and that a spanning tree of the entire network can be built using only measured lines.

4.7.1 Topological Observability

Topological observability as described in [9] is based on the condition that if a network is observable, then there exists at least one tree in the network that is also observable. It is therefore possible to find an observable tree by checking the observability and testing all the trees of the network. Determination of observability on a tree network is simple, but the number of trees grows rapidly with increase in network size. The IEEE 14-bus system has 3,909 trees. Since observability is equivalent to the existence of a spanning tree of full rank, a procedure is required to construct such a tree if it exists. Any connected collection of line flow measured branches is observable since the voltage difference across the line is easily computed from the flow through the line using a linearized model. In order to start building a spanning tree T of full rank, the starting point can be the collection of flow measured branches F. Since F may contain loops, these loops are pruned first. After the loops are removed, F may not be connected but may consist of several connected pieces, each of which is a tree. Such a nonconnected, loop-free subgraph is called a *forest.* In order to test observability, it is necessary to enlarge F to a spanning tree of full rank by adding branches to F.

In order to decide which branches must be added to enlarge F, an important simplification can be made. F is enlarged to a tree of full rank that is incident to every bus where there is no injection measurement. Such a tree is called a critical tree.

4.7.2 Topological Observability Algorithm

The algorithm for topological observability analysis can be described as follows: In the beginning the flow measurements are assigned to all the branches.

Next, assign injection measurements to reduce the existing forest by merging existing trees. After processing all the flows and injections, if a spanning tree cannot be found, then the observable islands are identified. This is achieved by discarding all injections that have at least one incident branch that is not part of branches forming a loop of the already defined forest. After updating the forest, search for more such injections until no more injections are to be removed.

Example 4.4

Determine maximal forest and observable islands for the 10-node network shown in Figure 4.6. The graph of the network is shown in Figure 4.7. Here, buses 3 and 4 are merged and numbered as 3. Buses 8, 9, and10 are merged and made bus 8.

Construction of the maximal forest starts with injection P_1, which can be assigned via P_{12} or P_{10}. We select edge P_{12}. Similarly, injection P_2 can be assigned via p_{21} or P_{23}, or P_{29}. Since P_{21} will form a loop with P_{12}, we select P_{29} Finally, we get a maximal forest, as shown in Figure 4.8.

It must be mentioned here that a maximal forest of full rank is not unique. However, the maximum number of observable buses, and therefore observable islands, is unique. The procedure for determining the observability is that assigned injections associated with a bus that has at least one incident branch not in the span of the maximal forest cannot be used in the solution. It therefore has to be removed from the forest. For the maximal forest shown in Figure 4.8, branches P_{46} and P_{56} do not form the loops with the branches defined by the forest of Figure 4.7. Thus the output of observability algorithm will indicate one observable island formed by buses 1, 2, 3, 4, 8, 9, and 10. Buses 5, 6, and 7 are unobservable.

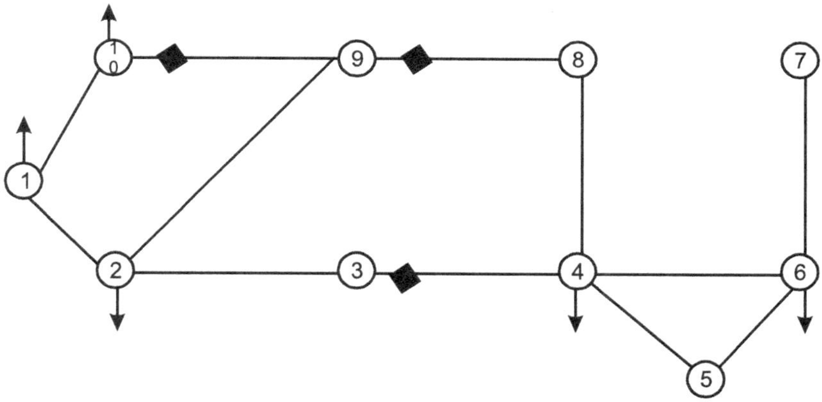

Figure 4.6 A 10-node network with measurements.

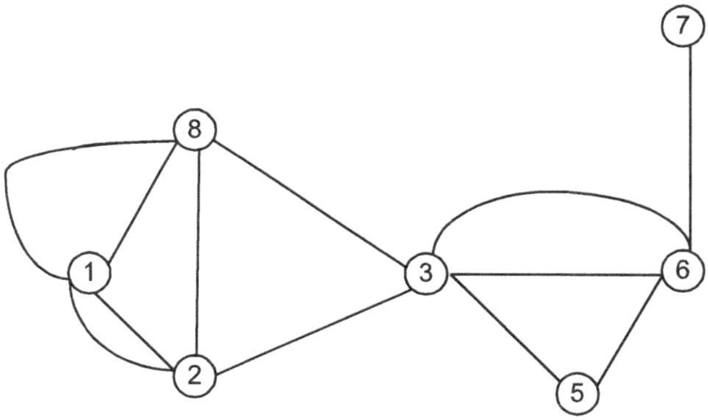

Figure 4.7 Graph of the 10-node network of Figure 4.6.

4.8 Topology Error Processing

A state estimation problem is formulated assuming that the network model and parameters are perfectly known. This model, known as branch to bus model, is built by topology processor by using information on the status of all circuit breakers and switching devices. Errors in this data can result in an error in state estimation. In Chapter 6, the detection of errors in parameters and measurement errors are discussed. Here we are more concerned about the errors in *topology*.

Topology errors can be broadly classified into two categories:

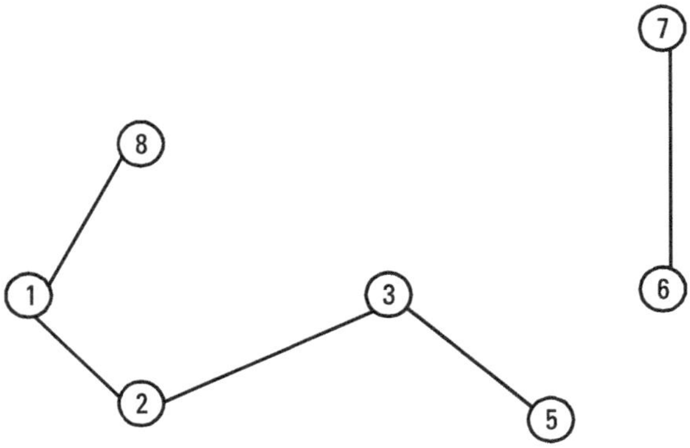

Figure 4.8 Maximal forest.

1. *Branch status errors* are errors affecting the status of regular network branches that either includes a disconnected element or excludes an energized element (line or transformer).
2. *Substation configuration errors* occur due to erroneous modeling of circuit breakers. Either a single electrical bus is modeled as two buses, or two buses are modeled as a single bus.

4.9 Detection and Identification of Topology Errors

Most of the work on detection and identification of topological errors is based on normalized residual tests. As discussed in the next section, topology errors can be detected using a two-stage approach:

- In the first stage, suspected buses are identified using normalized residuals of measurements.
- In the second stage, detailed models of suspected substations are introduced. The state estimation process is repeated using an expanded model of the system.

Most of the work on topology error detection and identification is based on normalized residual tests. The conditions of detectability and identifiabilty of topology errors are analyzed in detail in [11]. It has been shown that a single branch error cannot be detected if the branch is irrelevant (not having incident measurements) or the removal of the branch from the original network will cause the network to become unobservable (critical branch). A single branch error cannot be identified if it occurs in either one of critical pair of branches.

In general, adding additional extra measurements such as power flow measurements and power injection measurements can improve both detection and identification of topology errors. Recently the application of using PMUs in a few strategic places has been shown to greatly improve the topology error processing capability [13].

4.9.1 Residual Analysis

The branch status errors can be identified and corrected by means of either normalized residuals or state vector augmentation. The residual analysis makes use of results of converged state estimation to detect branch errors. The state estimator calculates an estimate state of the system based on the following equation:

$$z = h(x) + e \tag{4.32}$$

This equation is solved using an iterative procedure such as the linearized WLS problem of the following form:

$$z = H(x) + e \tag{4.33}$$

where $H(x)$ is the Jacobian matrix of $h(x)$

The effect of topology errors on the measurement equation can be modeled in the Jacobian matrix as

$$H = H_e + E \tag{4.34}$$

where H is the incorrect Jacobian matrix due to topology errors, and E is the Jacobian matrix error.

The measurement residual vector can be written as

$$r = z - H_e\hat{x} = (I - K_e)(Ex + e) \tag{4.35}$$

where is the identity matrix, and

$$K_e = H_e(H_e^T R^{-1} H_e)^{-1} H_e^T R^{-1} \tag{4.36}$$

If f is the vector of branch flow errors, and M is the measurement to branch incidence matrix, then the measurement bias error Ex can be written as a linear combination of errors in network branch flows as

$$Ex = Mf \tag{4.37}$$

and the residual vector will be given by

$$r = (I - K_e)Mf = Tf \tag{4.38}$$

It means a topology error in a certain branch j will produce a residual vector that is collinear with the jth column of T (T_j).

Thus branch j is single topology error detectable if $T_j \neq 0$. Also, branch j is single topology error identifiable if T_j is not collinear with any other column of T.

References

[1] Monticelli, A., and F. F. Wu, "Network Observability Theory," *IEEE Transactions on Power Apparatus and Systems,"* Vol. PAS-104, May 1985, pp. 1042–1048.

[2] Monticelli, A., and F. F. Wu, "Network Observability: Identification of Observable Islands and Measurement Placement," *IEEE Transactions on Power Apparatus and Systems,"* Vol. PAS-104, May 1985, pp. 1035–1041.

[3] Exposito, A. G., and A. Abur, "Generalized Observability Analysis and Measurement Classification," *IEEE Transactions on Power Systems,* Vol. 13, No. 3, August 1998, pp 1090–1095.

[4] Bose, A., and K. A. Clements, "Real Time Modeling of Power Networks," *Proceedings of the IEEE,* Vol. 75, No. 12, December 1987, pp. 1607–1622.

[5] Quintana, V. H., A Simoes-Costa, and A. Mandel, "Power System Observability Using a Direct Graph-Theoretic Approach," *IEEE Transactions on Power Apparatus and Systems,* Vol. PAS-101, March 1982, pp. 617–626.

[6] Clements, K. A., G. R. Krumpholz, and P. W. Davis, "Power System State Estimation with Measurement Deficiency: An Algorithm that Determines the Maximal Observable Subnetwork," *IEEE Transactions on Power Apparatus and Systems,* Vol. PAS-101, September 1982, pp. 3044–3052.

[7] Clements, K. A., G. R. Krumpholz, and P. W. Davis, "Power System State Estimation with Measurement Deficiency: An Observability/Measurement Placement Algorithm," *IEEE Transactions on Power Apparatus and Systems,* Vol. PAS-102, July 1983, pp. 2012–2020.

[8] Clements, K. A., and B. F. Woolenberg, "An Algorithm for Observability Determination in Power System State Estimation," paper A74 447-3 presented at the IEEE /PES summer meeting July 1975.

[9] Mori, H., and S. Tsuzuki, "A Fast Method for Topological Observability Analysis Using a Minimum Spanning Tree Technique," *IEEE Transactions on Power Systems,* Vol. 6, No. 2, 1991, pp. 491–500.

[10] Katsikas, P. J., and G. N. Korres, "Unified Observability Analysis and Measurement Placement in Generalized State Estimation," *IEEE Transactions on Power Systems,* Vol. 18, No. 1, 2003, pp. 324–333.

[11] Clements, K. A., and P. W. Davis, "Detection and Identification of Topological Errors in Electric Power System," *IEEE Transactions on Power Systems,* Vol. 3, No. 4, 1988, pp. 1248–1254.

[12] Chen, J., and A. Abur, "Enhanced Topology Error Processing Via Optimal Measurement Design," *IEEE Transactions on Power Systems,* Vol. 23, No. 3, 2008, pp. 845–853.

5

Bad Data Detection

5.1 Bad Data Detection in WLS Method

State estimation algorithms apply measured data obtained from the power system to a mathematical model to provide a reliable data base for monitoring, security assessment, and control functions. If the estimates obtained are not compatible with the standard deviations of quantities estimated, the possibility is that either the measured quantities are contaminated or the model is not accurate, or both. These algorithms may encounter four types of errors: measurement errors, wrong data, structural errors in the model, and parameter uncertainty in model parameters. One of the essential features of a state estimator is to detect measurement errors and identify and rectify them. Measurement errors may occur for many reasons. Random errors in measurement exist due to the finite accuracy of meters and communication medium. These errors can be filtered out by the state estimator, provided there is sufficient redundancy in measurements. The nature of filtering action depends on the specific method of estimation. Wrong data may be available due to defects in the meters, telecommunication failures, or noise caused by interference.

It is possible to identify some of the bad data easily and eliminated before filtering it by the estimator. For example, negative voltages, readings that are many times more or less than normal values, or large difference in incoming and outgoing values of currents at the nodes may be easily identified as bad data and are not considered in estimation. However, not all types of errors can be easily identified, and state estimators must be equipped with techniques to identify all types of bad data and filter them out.

Bad data detection and its processing depend on the method of state estimation in implementation. In this chapter, treatment of bad data for the WLS method of state estimation is discussed. If one or more gross errors are contained in the measurements the WLS state estimator may not reach a solution and may diverge. A well-known example when the WLS did not converge due to the existence of a topological error was an indirect contributing factor to the August blackout in the northeastern United States in 2003. The need to detect the gross error is a critical and challenging issue for WLS state estimation. Many algorithms have therefore been developed to detect gross errors in WLS state estimation.

In the WLS method of estimation the detection of bad data is done after the estimation process by processing the measurement residuals. Bad data is generally classified based on type, location, and number of measurements having error. They can be broadly classified as single bad data and multiple bad data. Multiple bad data is further classified as multiple noninteracting bad data and multiple interacting bad data. Single bad data means that there is only one measurement in the whole set that has a large measurement error [1–3].

Multiple bad data means there is more than one measurement that has large error. Multiple bad data may occur in measurements whose residuals are strongly or weakly correlated.

5.1.1 Leverage Points

The degree of interaction between measurements and analysis of errors can be carried out based on the sensitivities of measurement residuals to measurement errors. In power systems, leverage points are identified by examining the row vectors of the weighted Jacobian matrix. Those measurements whose associated row vector projections are distant from the bulk of all such projections are leverage points. Commonly, such measurements are associated with (1) power measurements on relatively short lines, (2) measured injections located on buses incident to a short line, and (3) measured injections located on buses with a relatively high number of incident lines.

A description of leverage points will be given in detail in Chapter 6 when dealing with robust state estimation.

5.2 Methods of Bad Data Detection

As stated earlier, there are various reasons for the data used for state estimation to be inaccurate. It is also known that WLS estimators are very sensitive to bad data. Detection of bad data refers to the determination whether the measurement set contains a bad data. Estimation in the presence of bad data can be performed indirectly in one of several ways. These methods can be classified as mathematical methods and intelligent methods. Among the mathematical methods are the chi-square distribution test, the largest normalized residual test, and hypothesis testing identification. Intelligent methods either require intensive training under different conditions or methods that do not require training but require high computation. These methods are therefore not very useful. Before removal of bad data it is important to detect the existence of bad data in the measurement set. Once the presence of bad data is detected, it is identified and removed or corrected.

5.2.1 Chi-Squares Test

The chi-square test is a convenient technique to detect the presence of bad data in measurement vector z. The objective function in the WLS state estimator is

$$\min J(X) = \sum_{i=1}^{m} \left(z_i - h_i(x)\right)^2 / R_{ii} \tag{5.1}$$

$$J(X) \sum_{i=1}^{m} (e_i)^2 / R_{ii} \tag{5.2}$$

$$= \sum_{i=1}^{m} \left(\frac{e_i}{\sqrt{R_{ii}}} \right)^2 \tag{5.3}$$

where e_i is the ith measurement error, R_{ii} is the diagonal entry of measurement error covariance matrix, and m is the number of measurements.

It is of the form:

$$Y = \sum_{i=1}^{N} \varkappa_i^2 \tag{5.4}$$

which corresponds to chi-squared $\varkappa^2(N)$ distribution, with N degree of freedom. In the power system state estimation the performance index J has

$(m\text{-}n)$ degrees of freedom, where m is the number of measurements and n is the number of state variables. This is because $m > n$, at most $(m - n)$ of the measurement errors will be linearly independent. A plot of $\chi^2(N)$ probability density function is shown in Figure 5.1, which presents the probability of finding $J(X)$ in the corresponding region.

In the chi-square testing method, if the value of $J(x)$ is more than a threshold value, it indicates the presence of bad data (see Appendix A) or the bad data is suspected, if

$$J\left(X > (m,n), p\right)$$

where p is the detection confidence level. Normally the value of p can be taken as 95%.

For choosing a threshold value tables are available in the statistical literature giving chi-square distribution function values for different degrees of freedom. The threshold is often chosen to constitute a 5% probability of error or false alarms. Once the existence of bad data is established, the next step is to identify the bad data in the set of measurements.

Example 5.1

A three-bus system shown in Figure 5.2 has the following data:

$$x_{13} = j0.01\ p.u.,\ x_{12} = j0.02,\ x_{23} = j0.01$$

The powers measured are

$$P_2 = -4.07\ p.u.,\ P_{31} = -1.90\ p.u.,\ \text{and}\ P_{13} = 2.04\ p.u.$$

The variances are

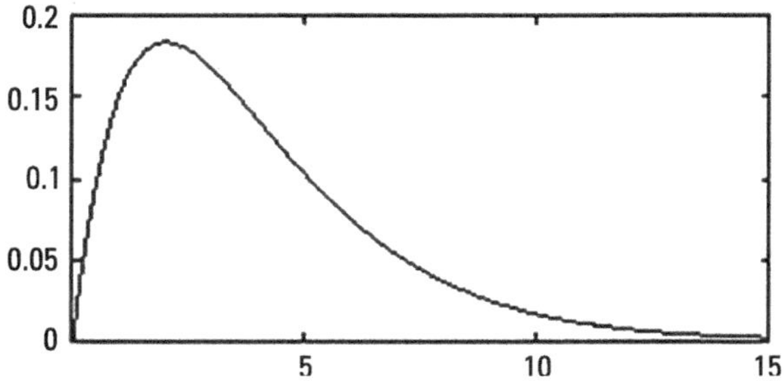

Figure 5.1 Chi-square distribution for four degrees of freedom.

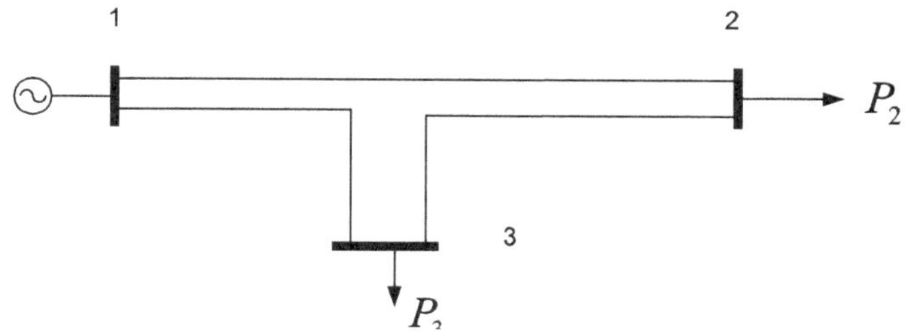

Figure 5.2 Three-bus system.

$$\sigma_{P_2}^2 = 0.004\ p.u.,\ \sigma_{P_{31}}^2 = 0.002\ p.u.,\ \sigma_{P_{13}}^2 = 0.002\ p.u.$$

Solution

From (5.2):

$$J(x) = \sum_{i=1}^{m} \left(\frac{e_i}{\sqrt{R_{ii}}} \right)^2$$

where R_{ii} is the diagonal entry of measurement error covariance matrix.

$$R_{ii} = \begin{bmatrix} \sigma_{P_2}^2 & & \\ & \sigma_{P_{13}}^2 & \\ & & \sigma_{P_{31}}^2 \end{bmatrix} = \begin{bmatrix} 0.004 & & \\ & 0.002 & \\ & & 0.002 \end{bmatrix}$$

The measurement vector estimate is given by $e_i = \left(z_i^{meas} - z_i^{est} \right)$. The objective function is obtained as

$$J(\hat{x}) = \sum_{i=1}^{m} \frac{\left(z_i - h_i(\hat{x}) \right)^2}{\sigma_i^2}$$

The estimated values are given by

$$\hat{x} = R_{z^{est}}\ W\ z$$

Here $R_{z^{est}}$ = covariance matrix, which is obtained as follows: First, the Jacobian matrix is obtained:

$$H = \begin{matrix} \\ P_2 \\ P_{13} \\ P_{31} \end{matrix} \begin{bmatrix} \theta_2 & \theta_3 \\ 150 & -100 \\ 0 & 100 \\ 0 & 100 \end{bmatrix}$$

The gain matrix $G^{-1} = [H'RH]^{-1} = 10^{-7} \begin{bmatrix} 2.222 & 0.667 \\ 0.667 & 1.00 \end{bmatrix}$

The covariance matrix of measurement estimate is given as

$$R_{z^{est}} = HRH' = \begin{matrix} P_2 \\ P_{13} \\ P_{31} \end{matrix} 10^{-3} \begin{bmatrix} 4 & 0 & 0 \\ 0 & 1 & -1 \\ 0 & -1 & 1 \end{bmatrix}$$

The measurement vector estimate is

$$z^{est} = R_{z^{est}} W z^{meas}$$

$$= 10^{-3} \begin{bmatrix} 4 & 0 & 0 \\ 0 & 1 & -1 \\ 0 & -1 & 1 \end{bmatrix} \begin{bmatrix} 250 & & \\ & 500 & \\ & & 500 \end{bmatrix} \begin{bmatrix} -4.07 \\ 2.04 \\ -1.90 \end{bmatrix} = \begin{bmatrix} -4.07 \\ 1.97 \\ -1.97 \end{bmatrix}$$

$$e_i = \left(z_i^{meas} - z_i^{est}\right) \begin{bmatrix} 0.00 \\ 0.07 \\ 0.07 \end{bmatrix}$$

$$\sum e_i^2 = 0 + 0.0049/0.004 + 0.0049/0.004 = 2.5$$

Here, $m = 3$ and $n = 2$, degree of freedom = 1.

Now the lookup table can be used to find the test threshold for 95% confidence level and degree of freedom =1 which is 3.841. Since $J(x) = 2.5$ is less than 3.841, no bad data will be suspected.

5.3 Identification of Bad Data

Chi-square test is used to detect only the presence of bad data in set of measurements. There are methods that detect and also identify the bad data so that these can be removed. These methods are described here.

5.3.1 Method of Normalized Residual

In WLS state estimation, bad data processing occurs after convergence by processing the measurement residuals and the associated covariance matrix. The difference between the estimate and the measurement z_i is called the measurement residual r_i. The measurement residual equations can be obtained starting from linearized measurement equation

$$\Delta z = H\Delta x + e \tag{5.5}$$

The following assumptions are made regarding the error e:

- Errors are assumed to have normal distribution with zero mean and known standard deviation σ_i or $E(e_i) = 0$.
- The errors are assumed to be independent or $E(e_i e_j) = 0$. Hence the covariance matrix of the measurement error is diagonal.

Applying the optimization criterion, the following expression can be derived for the optimal state update.

$$\Delta\hat{x} = \left(H^T R^{-1} H\right)^{-1} H^T R^{-1} z = G^{-1} H^T R^{-1} \Delta z \tag{5.6}$$

The calculated measurement updates based on the estimated state updates will be given by

$$\Delta\hat{z} = H\Delta\hat{x} = HG^{-1}H^T R^{-1} z = K\Delta z \tag{5.7}$$

where $K = HG^{-1}H^T R^{-1}$ is called the hat matrix. The matrix K has the following property: K.K.K....K = K and K.H = H and (1-K) H = 0.

The measurement residual is the difference between the estimated value and the measured value of the data. The expression for measurement residuals can be derived as follows:

$$\begin{aligned} r &= \Delta z - \Delta\hat{z} \\ &= (I - K)\Delta z \end{aligned} \tag{5.8}$$

$$= (I - K)(H\Delta x + e) = (I - K)e \tag{5.9}$$

$$= Se \tag{5.10}$$

where S is called *the residual sensitivity matrix.* It represents the sensitivity of measurement residuals to measurement errors. It has the following properties:

$$S \cdot S \cdot S \ldots S = S \tag{5.11}$$

5.3.2 Normalized Residuals

The probability density function for residual r is also a normal distribution function having zero mean and a standard deviation σ. If we divide the measurement residual r by the standard deviation σ then we get the *normalized measurement residuals.* The normalized measurement residual is designated by r_i^{norm}, which also has a normal probability distribution and has a standard deviation of unity. It is possible to decide on the presence of bad data by comparing the largest value of r_i^{norm} against a threshold value. The threshold value can be selected statistically based on the level of sensitivity desired.

The property of normalized residuals can be used to identify the bad measurement data. However, the properties of measurement will affect the process of bad data detection. In a power system, the measurements are power active and reactive, voltage, and currents, which are spread over a wide area. These measurements may be critical or redundant. A critical measurement is one whose elimination from measurement may make the system unobservable. Redundant measurement is a measurement that is not critical. The redundant measurements can have their nonzero residuals. However, the measurement residual of critical measurements will always be zero. If there are two redundant measurements whose simultaneous removal from the measurement set can make the system unobservable, then these measurements are known as a *critical pair.*

It can be shown that if there is only one bad data in the measurement set that neither belongs to critical measurement nor a member of the critical pair, then the largest normalized residual will correspond to the bad measurement. Note that the detection and identification process can be performed simultaneously bypassing the chi-square test and directly computing the normalized residuals. However, the computation cost of calculating normalized residuals is large and thus the use of a chi-square test for preliminary identification of bad data may be justified.

5.3.3 Largest Normalized Residual Test

In this method, normalized residuals are used to identify the bad data. The largest normalized residual can be obtained in the following simple manner.

First, determine the normalized residual by the measurement variances as given by

$$r_i = \left[z_i - h_i(v,\theta)\right]/\sigma_{i\,i} \tag{5.12}$$

where r_i is the normalized residual. Now rank the residuals and eliminate the residuals above some threshold or simply the largest one. This is known as the *largest normalized residual* (r^N_{max}) test. Now the estimation problem is repeated without the largest residual. The normalized vector of residuals is obtained as follows:

1. Solve the WLS estimation and obtain the elements of measurement residual vector r_i.
2. Compute the normalized residuals:

$$r_i^{norm} = \frac{|r_i|}{\sigma_i} \, i = 1 \ldots m$$

3. Find the largest normalized residual.
4. Eliminate this measurement from the measurement set and then again compute the normalized residuals.

In eliminating the bad data actual removal of bad measurement from the set is avoided as shown below. Assume that all are error-free except the measurement z_i, which can be written as

$$z_i^{meas} = z_i^{true} + b_i\sigma_i \tag{5.13}$$

where b_i is the magnitude of bad data in terms of standard deviation σ_i.

Using the linearized sensitivity relation, the bad measurement residual can be written as

$$r_i^{bad} \cong S_{ii}e_i = z_i^{bad} - h(\hat{x}) \tag{5.14}$$

where $h(\hat{x})$ is the state estimate based on the measurement set that also includes the bad measurement. A simple three-bus system is shown in Example 5.2 to demonstrate the detection and identification of bad data.

Example 5.2

A three-bus sample system is shown in Figure 5.3. The following set of measurements are available: $P_1 = 3.90\ p.u.$, $P_2 = -4.05\ p.u.$, $P_3 = -0.48\ p.u.$, $P_{13} = 2.04\ p.u.$

The variances are $\sigma_1^2 = 0.001$, $\sigma_2^2 = 0.004$, $\sigma_3^2 = 0.001$, $\sigma_{13}^2 = 0.002$

Solution

The Jacobian $H = \begin{bmatrix} -50 & -100 \\ 150 & -100 \\ -100 & 200 \\ 0 & -100 \end{bmatrix}$

The covariance matrix of measurement vector is

$$R = \begin{bmatrix} 0.001 & & & \\ & 0.004 & & \\ & & 0.001 & \\ & & & 0.002 \end{bmatrix}$$

The covariance matrix of state estimate is

$$R_x = G^{-1} = (H'R^{-1}H)^{-1}$$

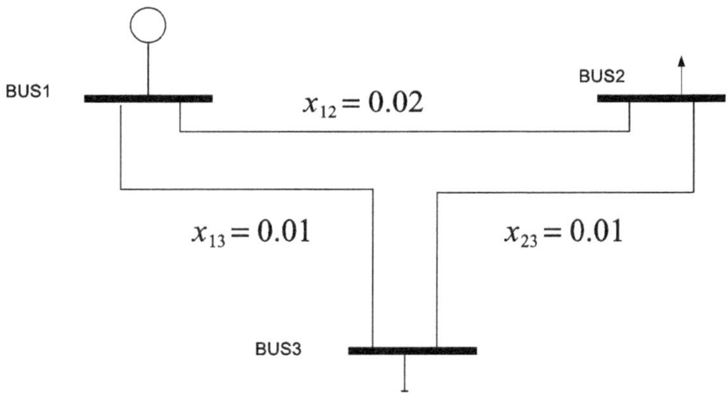

Figure 5.3 Three-bus system for Example 5.2.

The gain matrix = G

$$= \begin{bmatrix} -50 & 150 & -100 & 0 \\ -100 & -100 & 200 & -100 \end{bmatrix} \begin{bmatrix} 1000 & & & \\ & 250 & & \\ & & 1000 & \\ & & & 500 \end{bmatrix} \begin{bmatrix} -50 & -100 \\ 150 & -100 \\ -100 & 200 \\ 0 & -100 \end{bmatrix}$$

$$= \begin{bmatrix} 18125000 & -18750000 \\ -18750000 & 57500000 \end{bmatrix}$$

$$G^{-1} = 10^{-6} \begin{bmatrix} 0.0832 & 0.0271 \\ 0.0271 & 0.0263 \end{bmatrix}$$

The state estimate $x_{est} = G^{-1}H'Wz = \begin{bmatrix} -0.400 \\ -0.0201 \end{bmatrix}$.

The estimated values of measured quantities z_{est} are

$$z_{est} = R_z Wz$$

The covariance matrix of measurement estimate R_z is

$$HG^{-1}H' = 10^{-4} \begin{bmatrix} 7.42 & -6.52 & -1.10 & 3.98 \\ -6.32 & 13.22 & -6.90 & -1.43 \\ -1.10 & -6.90 & 8.00 & -2.55 \\ 3.98 & -1.43 & -2.55 & 2.63 \end{bmatrix}$$

Covariance matrix of residuals $R_r =$

$$z_{est} = 10^{-4}\begin{bmatrix} 7.42 & -6.52 & -1.10 & 3.98 \\ -6.32 & 13.22 & -6.90 & -1.43 \\ -1.10 & -6.90 & 8.00 & -2.55 \\ 3.98 & -1.43 & -2.55 & 2.63 \end{bmatrix}\begin{bmatrix} 1000 & & & \\ & 250 & & \\ & & 1000 & \\ & & & 500 \end{bmatrix}\begin{bmatrix} 3.90 \\ -4.05 \\ -048 \\ 2.04 \end{bmatrix}$$

$$= \begin{bmatrix} 3.992 \\ -3.61 \\ -0.374 \\ 2.09 \end{bmatrix}$$

$$r_i = \begin{bmatrix} -0.092 \\ -0.44 \\ -0.106 \\ -.05 \end{bmatrix}$$

and normalized residuals are

$$r_n = \begin{bmatrix} 2.90 \\ 6.95 \\ 3.16 \\ 1.11 \end{bmatrix}$$

The measurement P_2 is having the largest normalized residual. If the maximum value is assumed for normalized residual to be 3, then measurement P_2 and P_3 represents bad data.

5.4 Hypothesis Testing Identification

The HTI method was first introduced by L. Mili in 1984 [4–6]. It uses the properties of measurement residuals to find all the bad measurements in one iteration. This method is based on two hypotheses, the *null hypothesis* denoted by H_0 and the *alternative hypothesis* denoted by H_1.

Here H_0 is considered as no bad data or structural error present.

H_1 means that H_0 is not true or that bad data is present.

The HTI method makes use of normalized residuals to determine the reduced form of sensitivity matrix S. First the WLS state estimator is run and normalized residuals are calculated. A set of measurements that are noncritical and linearly independent with largest normalized residuals is selected, and the rest of the measurements are assumed to be free from error. Then the sensitivity matrix and error covariance matrix are partitioned as suspect and true measurements.

For example, if two out of six measurements have normalized residuals more than the threshold value then the hypotheses can be stated as

- Measurement 1 is good and measurement 2 is bad;
- Measurement 2 is good and measurement 1 is bad;
- Both the measurements are bad.

Two types of errors can occur:

1. Error I: If P_e denotes the false alarm probability (i.e., P_e is the probability of rejecting H_0 when H_0 is actually true). It means that the ith measurement has no bad data but it is identified as bad data. Normally the value of P_e is small; typical values are 0.01, 0.05, or 0.10. For larger values, the results will be sensitive to random errors.
2. Error II: Also, if P_d is the probability of detection of H_1 as true when it is actually true. It means that the ith measurement has bad data and is identified as bad data. If it is assumed that the probability distribution of these errors is normalized, then mean value of error I is zero. The mean value of error II is e.

Example 5.3

Consider the three-bus problem as given in Example 5.2. The largest normalized residuals are found. From the chi-square test for 2.5% of false alarm, $\alpha = 0.025$, degree of freedom = 1, and the critical value is 5.024.

The largest normalized residual found is 6.95, hence the NULL hypothesis is neglected.

5.5 Case Study: Improved Bad Data Processing with Strategic Placement of PMUs

The application of synchronized phasor measurement units is gradually increasing in power systems. Since the key issue in detecting the bad data is measurement redundancy, the number of data can be increased by incorporating the synchronized phasor measurements into state estimators along with conventional measurements. It is therefore possible to transform certain critical measurements indo redundant measurements by adding data from PMUs placed at proper locations [8].

In principle, any additional measurement added improves the redundancy. However, the addition of a PMU can be more useful as it can provide the voltage phasor at the bus and phasor currents on several or all lines incident to that bus. In cases where the power system is fully observable from conventional measurements, installation of a few PMUs at suitable locations can make some of the critical measurements redundant.

Here the optimal placement of PMUs from the point of view of improving the bad data detection and processing is discussed. First, optimal placement of PMUs in order to convert all critical measurements into redundant ones is considered. The system with these measurements added will be no more vulnerable to the loss of any of these critical measurements. It is therefore important to determine the critical measurements in the set and then find the placement locations for PMUs to convert all critical measurements into a redundant one.

A critical measurement is a measurement whose elimination from the measurement set will make the system unobservable. A power system will be observable only if the Jacobian measurement is of full rank. For an n-bus power system with m measurements, the Jacobian measurement is an $m \times (n - 1)$ matrix with a column rank of $(n - 1)$. The column rank is the maximum number of its linearly independent columns. The critical measurements can be identified using observability analysis. Here a numerical method for identification of critical measurements is described.

1. Form the Jacobian measurement matrix H.
2. Augment H by adding all candidate PMUs at the bottom. Let the submatrix representing the rows corresponding to candidate PMUs be denoted by H_c.
3. Perform a rectangular lower trapezoidal and upper (LU) factorization on H. This can be accomplished by using the Peters-Wilkinson method [7]. Row pivoting in this factorization should be limited to existing measurements only and no PMU measurement is disturbed. If such pivoting proves inadequate, then it implies an unobservable system. In that case row pivoting will also be extended to candidate measurement so that the system becomes observable.

4. Now after pivoting of the rows. the top n rows of H will represent the minimally observable set of measurements. Let the H matrix after LU decomposition be expressed as

$$H = \begin{bmatrix} L_0 \\ M_r \\ M_c \end{bmatrix} \cdot [U] \tag{5.15}$$

where L_0 and U are lower and upper triangular factors of dimension (n–10). Rows of L_0, M_r, and M_c correspond to the essential measurements, redundant measurements, and the candidate PMU measurements, respectively.

Now calculate the test matrix T as

$$T = \begin{bmatrix} T_r \\ T_c \end{bmatrix} = \begin{bmatrix} M_r \\ M_c \end{bmatrix} [L_0]^{-1} \tag{5.16}$$

If column k of T_r contains all zero elements then it will correspond to critical measurement. Then the rows having nonzero entries in column k of T_c will correspond to the candidates that will transform critical measurement into a redundant one.

Example 5.3: Determination of Optimum PMU Placement

A six-bus power system is shown in Figure 5.4. Consider the existing measurements as shown in the figure. The following matrices will be obtained:

$$H_0 = \begin{bmatrix} 1 & 0 & 0 & 0 & 0 & 0 \\ -1 & 3 & -1 & 0 & -1 & 0 \\ 0 & 0 & -1 & 0 & -1 & 2 \\ 0 & 0 & 0 & 1 & -1 & 0 \\ 0 & 0 & 1 & 0 & -1 & 0 \\ 1 & -1 & 0 & 0 & 0 & 0 \end{bmatrix}$$

$$H_r = \begin{bmatrix} 0 & 1 & 0 & 0 & -1 & 0 \\ 0 & -1 & 1 & 0 & 0 & 0 \end{bmatrix}$$

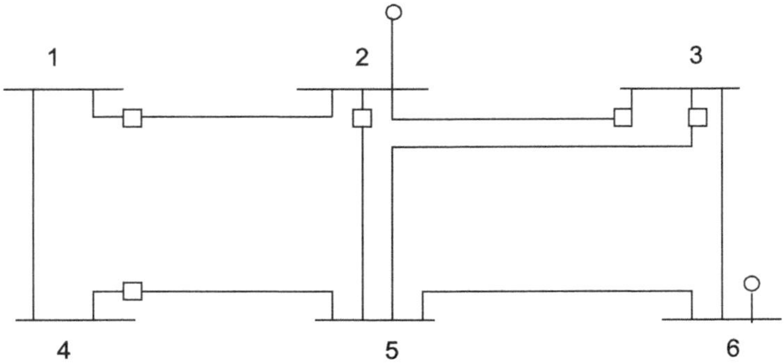

Figure 5.4 A six-bus power system.

Assuming PMU at every bus:

$$H_c = \begin{bmatrix} 0 & 1 & 0 & 0 & 0 & 0 \\ 0 & 0 & 1 & 0 & 0 & 0 \\ 0 & 0 & 0 & 1 & 0 & 0 \\ 0 & 0 & 0 & 0 & 1 & 0 \\ 0 & 0 & 0 & 0 & 0 & 1 \end{bmatrix}$$

After bifurcation the matrices are

$$L_0 = \begin{bmatrix} 1 & 0 & 0 & 0 & 0 & 0 \\ -1 & 1 & 0 & 0 & 0 & 0 \\ 0 & 0 & 1 & 0 & 0 & 0 \\ 0 & 0 & 0 & 1 & 0 & 0 \\ 0 & 0 & -1 & 0 & 1 & 0 \\ 1 & -0.33 & -0.33 & 0 & 0 & 1 \end{bmatrix}$$

$$M_r = \begin{bmatrix} 0 & 0.33 & -0.33 & 0 & 0.5 & 0.5 \\ 0 & -0.33 & -0.66 & 0 & 0.5 & -0.5 \end{bmatrix}$$

$$M_c = \begin{bmatrix} 0 & 0.33 & -0.33 & 0 & 0 & -1 \\ 0 & 0 & 1 & 0 & 0.5 & -1.5 \\ 0 & 0 & 0 & 1 & -0.5 & -1.5 \\ 0 & 0 & 0 & 0 & -0.5 & -1.5 \\ 0 & 0 & 0 & 0 & 0 & -1.5 \end{bmatrix}$$

The test matrix $T = \begin{bmatrix} T_r \\ T_c \end{bmatrix}$ where

$$T_r = \begin{bmatrix} 0 & 0.5 & 0 & 0 & 0.5 & 0.5 \\ 0 & -0.5 & 0 & 0 & 0.5 & -0.5 \end{bmatrix}$$

and

$$T_c = \begin{bmatrix} 1 & 0 & 0 & 0 & 0 & -1 \\ 1 & -0.5 & 0 & 0 & 0.5 & -1.5 \\ 1 & -0.5 & 0 & 1 & -0.5 & -1.5 \\ 1 & -0.5 & 0 & 0 & -0.5 & -1.5 \\ 1 & -0.5 & 0.5 & 0 & 0 & -1.5 \end{bmatrix}$$

Checking the columns of T_r, it is found that column 1, 3, and 4 have zero elements. Thus the measurements corresponding bus 1, 3, and 4 qualify for critical measurements. However, the column corresponds to phase angle measurement at bus 1, and can be taken as reference bus. Thus only the power injection measurement at bus 6 and power flow measurement from bus 4 to bus 5 are the critical measurements, as they correspond to columns 3 and 4.

Now the matrix T_c will be considered to identify the candidates that can transform these critical measurements into redundant one. Here bus 3 has P_{32} and P_{35} power flow measurements whereas bus 4 has P_{45} phase angle measurements at bus 3. Therefore bus 4 will be identified as candidate measurements.

Now the final step involves the optimal selection of the minimum number of required PMUs from the list of candidates identified above. This will be an optimization problem where an optimal subset of identified candidate measurements will be determined. This will require another integer programming problem as given below:

$$\text{Minimize} \sum_{i}^{n} c_i x_i \tag{5.17}$$

where

n is equal to number of buses in the system

C_i = cost of installation of PMU at bus

Subject to the condition $F \cdot X \geq \hat{1}$

where x is a binary array whose entries correspond to candidate measurements. In this example it is the phase angle measurements at buses 2 to 6. The entries will be either 1 or 0 depending on whether it is selected or not, respectively.

F is the critical measurement to candidate measurement incidence matrix

$F_{ij} = 1$, if j is candidate for critical measurement i, otherwise it is 0

$$\hat{1} = [1.1.1....1]^T$$

In above example, there are only two nonoverlapping measurements, hence both of these candidates will be selected.

References

[1] Clements, K. A., and P. W. Davis, "Multiple Bad Data Delectability and Identifiabilty: A Geometric Approach," *IEEE Transactions on Power Delivery*, Vol. 1, No. 3, 1986, pp. 355–360.

[2] Handschein, E., F. C. Schweppe, and J. Kohlas, et al., "Bad Data Analysis for Power System State Estimation," *IEEE Transactions on Power Apparatus and Systems,* Vol. 94, No. 2, 1975, pp. 329–337.

[3] Van Cutsem, T., M. Ribbens-Pavella, and L. Mili, "Bad Data Identification Methods in Power System State Estimation: A Comparative Study," *IEEE Transactions on Power Apparatus and Systems*, Vol. PAS-104, No. 11, November 1985, pp 3037–3049.

[4] Van Cutsem, T., M. Ribbens-Pavella, and L. Mili, "Hypothesis Testing Identification: A New Method for Bad Data Analysis in Power System State Estimation," *IEEE Transactions on Power Apparatus and Systems,* Vol. PAS-103, No. 11, November 1984, pp. 3239–3252.

[5] Petes, G., and J. H. Wilkinson, "The Least Squares Problem and Pseudo Inverses," *Computer Journal,* Vol. 13, No. 3, 1970, pp. 309–316.

[6] Merril, H. M., and F. C. Shweppe, "Bad Data Suppression in Power System Static State Estimation," *IEEE Transactions on Power Apparatus and Systems,* Vol. PAS 90, 1971, pp. 2718–2725.

[7] Zhuang, F., and R. Balasubramanyam, "Bad Data Processing in Power System State Estimation by Direct Data Detection and Hypotheses Tests," *IEEE Transactions on Power Systems,* Vol. 2, No. 2, 1987 pp. 321–327.

[8] Chen, J., and A. Abur, "Improved Bad Data Processing Via Strategic Placement of PMUs," *IEEE Power Engineering Society General Meeting,* 2005, pp. 509–513.

Appendix 5A: Chi-Square Test

Chi-square is a statistical test normally used to determine if a sample of data came from a population with a specific distribution. Consider a set of k measurements $\{x_1, x_2, \dots x_k\}$ and if these measurements have normal distribution with mean μ and standard deviation σ, a chi-square test can be formulated as

$$\mathcal{X}^2 = \sum_{i=1}^{k} \frac{(x_i - \mu)}{\sigma^2} \tag{5A.1}$$

The most common application of a chi-square test is to compare observed values of data with an expected value. There are tables available for calculating chi-square value for different degree of freedom. Using this test, cumulative probability can be obtained.

6

Robust State Estimation

6.1 Basic Formulation

In a power system, estimators are used to provide a reliable estimate of unknown states in a given mathematical model from available redundant measurements. The measurement set consists of

- *Telemetered measurements*, which are online telemetered bus voltage magnitudes, active and reactive power flows, active and reactive injections (subjected to noise or error in metering, communication system, and so forth).
- *Pseudomeasurements* or zero injections, which contain no error.

The state estimation programs are also built with an ability to clean the erroneous data that may be contaminated due to various factors. If the measurements have normally distributed errors, WLS method, which is the most common method used in the industry for state estimation, provides an optimum solution to the state estimation problem. However, if some of the measurements have very large errors (statistical outliers), then the WLS method becomes highly unreliable. Since the works of Adibi and Thorne [1] and Adibi and Stoval [2], there is a growing concern about the poor quality of measurements provided by the SCADA system. The measurements in a conventional SCADA system have two main sources of errors that degrade the quality of data. The first one of these is error due to calibration of instruments, as these are calibrated at best every two years. Thus, the accuracy of the instruments deteriorates with time due to variation in temperature and environmental effects that may not be compensated. The second source of bias stems from the scaling procedure (for stepping down the current and voltage) currently in vogue in various control centers. Since scaling is a linear procedure, it is not able to account for errors in nonlinear equipments such as instrument transformers. As a result, the power measurements are strongly biased. Such measurements are called outliers. Other sources of errors are measurement time skews, unbalancing in three-phase systems, incorrect wiring, instrument failures, parameter and topology errors, and so forth. In Chapter 5, methods for bad data detection and elimination were discussed. In this chapter, estimation methods that are robust (not affected by bad data) are described.

Due to the poor quality of data provided by the SCADA system, the main objective of state estimation is to minimize the bias in the database of an energy management system. Bias is defined as the difference between the measured value of any variable and its unknown "true value." Fortunately, bias analysis of estimators has long been a central issue in robust statistics. There is well-developed theory available in statistics to deal with the problem of bias. Applying these statistical theories various robust estimators for power system have been developed. These estimators, known as robust estimators, try to minimize this problem by processing the measurements so that outliers have little or no effect on the estimated states. Specialized algorithms have been developed to solve estimation problems with corrupted measurements. Some commonly used algorithms are discussed in this chapter.

6.2 Breakdown Points

Since the initial proposal made by Hodges and further developed by Hampel, the concept of breakdown point has proved to be a powerful tool for the analysis and design of robust estimators [3, 4]. Roughly speaking, the breakdown point is defined as the smallest fraction of outliers that can cause an estimator to take arbitrary large values and ruin an estimator. The outliers are those measurement data that have abnormally different values from rest of the set of measurements. These outliers are present in the set due to gross errors in the measurement. The degree of robustness of an estimator in the presence of outliers can be measured by the concept of the breakdown point. The finite sample breakdown point is the measure of maximum fraction of outliers that a given sample may contain without spoiling the estimate completely. Thus the larger the breakdown point an estimator has, the more robust it is. A finite sample version of the breakdown point that has gained wide acceptance was formulated by Donoho and Huber [5]. It has been called the exact fit point to indicate that the majority of measurements are free of errors, and lie exactly on a hyperplane when linear regression models are considered. In statistics, linear regression is an approach to model the relationship between two variables by fitting a linear equation to observed data. One of these variables is a dependent variable and the other is an explanatory variable. If there is only one explanatory variable it is called simple regression. If more than one explanatory variable is present it is known as multiple regression.

In multiple linear regression, each of m measurements z_i (dependent variable) are represented in terms of n unknown state variables x_i and n explanatory variables $l_{i1} \ldots l_{in}$

In equation form it can be written as

$$z_i = l_{i1}x_1 + \ldots + l_{in}x_n + e_i \ldots i = 1,2,\ldots m \tag{6.1}$$

or

$$z = Hx + e \tag{6.2}$$

Here H is the $m \times n$ matrix, known as the design matrix of regression.

$$H = \begin{bmatrix} l_{11} & l_{12} & \cdots & l_{1n} \\ l_{21} & l_{22} & \cdots & l_{2n} \\ \vdots & \vdots & \cdots & \vdots \\ l_{n1} & \cdots & \cdots & l_{nn} \end{bmatrix} \tag{6.3}$$

The row vector of H, $l_i^T = l_{i1} \ldots l_{in}$, defines a point in n-dimensional subspace, called the factor space of regression.

In a power system, the linear regression model is obtained with DC approximations. It is derived by linearizing the n-bus system model about the flat voltage profile and neglecting the series resistances and shunt capacitances of the branches. Two linear models as proposed in [2] can be written. One model relates the real power measurements to the (n-1) voltage phase angles θ_i. The other model relates to the reactive power measurements P_i to the n nodal voltages magnitudes. Then H is the P, θ or Q, V Jacobian submatrix, respectively.

The robustness of an estimator can be quantified by a finite sample breakdown point. Consider a set Z of m good measurement

$$Z = \{(z_1, l_1), (z_{2,}, l_2), \ldots (z_m, l_m)\} \tag{6.4}$$

The regression estimator T maps Z into the state estimate

$$x = \widehat{T(z)} \tag{6.5}$$

Assuming that out of m good measurements there are m_b data points that are outliers, if we now replace good data points by outliers, it yields a contaminated set Z'. This results in a new states vector $\hat{x}'$. The maximum bias in the estimate will be

$$b^{m_b} = \max\{\hat{x}_i - \hat{x}_i'\} \; i = 1, \ldots, n \tag{6.6}$$

where b^{mb} is the difference or bias in the estimated value when a number of good data points are replaced by outliers.

Here m_b is the number of bad data or outliers. Starting with one by one number of bad data points out of total number m_b, the number is increased until the value of b^{m_b} is no longer finite. The largest ratio of bad data m_b to total data m for which b^{m_b} remains bounded is known as the breakdown point of that estimator. A robust estimator has a large breakdown point, which means it can handle a larger number of bad data and still produce correct estimates.

In the case of linear regression models, robustness can be specified by another simpler concept known as exact fit property [7]. In the linear regression model, an estimator of n unknown state variables is said to have an exact fit property if the estimate is an n-dimensional hyperplane and majority (more than 50%) of the measurements lie on this hyperplane. Thus the exact fit point of an estimator is the maximum fraction of outliers for which exact fit property is true.

Example 6.1

Determine the breakdown point for the following set of voltage measurement observations:

$$Z = 0.98, 1.00, 1.05, 1.07, 1.10$$

Solution

If the estimated value is obtained as the average of measured values

$$\hat{Z} = \{0.98 + 1.00 + 1.05 + 1.07 + 1.10\}/5 = 1.04$$

If one of the measurements is an outlier, say measurement 1.10 is replaced by an arbitrarily large value of ∞, then the estimated value becomes ∞. This means that even one abnormally large value will ruin the estimator, or the breakdown point for this estimator is zero.

If a median estimator is used instead of a mean estimator, then the estimated value is 1.05. Now if value 1.10 is replaced by ∞, the estimated value is still 1.05. If two measurements 1.07, 1.10 are replaced by ∞ each, the estimated value is still 1.05. However, if three measurements 1.05, 1.07, and 1.10 are replaced by ∞, the estimated value also becomes ∞ (unbounded). Thus the ratio 2/5 = 0.4 is the breakdown point and the median estimator is more robust than the mean estimator.

6.2.1 Leverage Points

Since the presence of even a single bad data will distort the WLS estimate, improvement in bad data detection and estimation methods have been developed for power system state estimators. Among these methods, the *weighted least absolute value estimator* (WLAV) that minimizes the weighted sum of absolute errors has been found to be simpler. The reason is that WLAV estimation problem can be solved as linear programming problem (Appendix 6A). WLAV estimators successfully reject bad measurement data as long as none of these measurements are leverage points [6]. Leverage points are defined as the points of a regression that are far away from the bulk of the data points in the factor space (Figure 6.1).

A common practice is to flag an observation as having a large leverage value, if its leverage value is three times larger than the mean leverage value [7]. Although the characterization of a leverage point depends only on the independent variable x, their classification as good or bad data also depends on the measured values z as well as on the corresponding variances. The location of points in *x-space* is important in determining the property of a regression model. Remote points in *x-space* have a large impact on parameter estimates, standard error, and model summary. There are two approaches to solve the problem when a measurement data set has leverage points:

1. To identify the leverage points and eliminate them before estimating.
2. To use estimators that are not sensitive to leverage points.

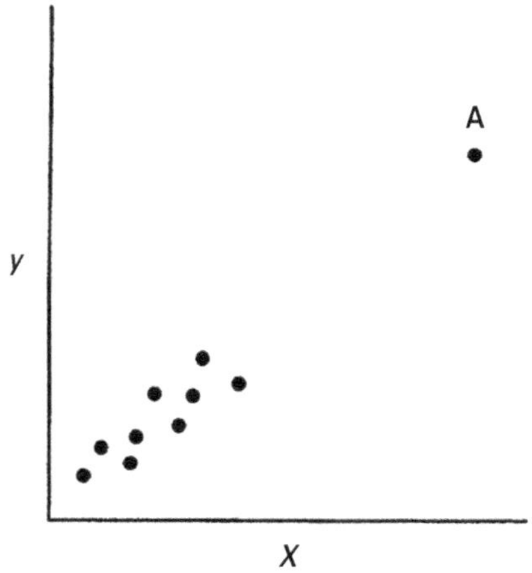

Figure 6.1 An example of leverage points.

A leverage point can be identified considering the linear regression model given below

$$z = Hx + e \tag{6.7}$$

where $(e) = 0$, $E[e, e^T] = I_m$ and e is m dimensional state vector that accounts for the uncertainty in the measurements and models.

The least square estimators for variables x and z are given by

$$\hat{x} = \left[H^T H\right]^{-1} H^T z \tag{6.8}$$

and

$$\begin{aligned} \hat{z} &= H\left[H^T H\right]^{(-1)} H^T z \\ &= Kz \end{aligned} \tag{6.9}$$

where $K=H[H^TH]^{-1}H^T$ is the hat matrix defined in Chapter 3. The diagonal elements of k can be expressed as

$$K_{ii} = H_i\left[H^T H\right]^{-1} H_i^T \tag{6.10}$$

where H_i denotes the ith row of matrix H. The value of K_{ii} indicates the measure of the distance of measurement factor H_i from the bulk of remaining (m-1) measurement factors. The residuals of measurement are related to the elements of hat matrix K as

$$r = z - \hat{z} = z - Kz \tag{6.11}$$

from (6.11) it is clear that the residuals of measurement corresponding to a leverage point will be very small (value of K is large) even though it may be contaminated by a large error. Hence we can say that the WLAV estimator loses its robustness when leverage points are present. That is why the identification and removal of the effect of leverage point is an active area of research. The combinational approach and least median of squares (LMS) method discussed later in this chapter are better methods to detect bad data in the presence of leverage points.

Leverage measurements normally occur in connection with low impedance branches and nodal injection measurements and also where injections in nodes with large number of branches are present. In order to avoid the occurrence of certain types of leverage points, low impedance branches with the corresponding through flows are considered as additional state variables.

6.3 M-Estimators

A number of robust state estimators have been developed that can work even in the presence of outliers. Among them are the M-estimator (maximum likelihood type estimator), L-estimator (linear combination of order statistics), R-estimators (estimator based on rank transformation), and LMS estimators.

M-estimation in the context of regression was first introduced by Huber (1973) [5] as a result of making the least squares approach robust. Although M-estimators are not very useful for measurements with leverage points, they are popular in applications where leverage points are not an issue. Instead of minimizing a sum of squares of the residuals, in general an M-estimator minimizes an objective function that is expressed as a function of measurement residuals, subject to constraints given by measurement equations. There are several M-estimators available in the literature. Here only few simple M-estimators used in power system state estimation are discussed, starting with the classical linear regression model:

$$z = h(x) + r \tag{6.12}$$

minimize

$$\text{minimize } J(x) = \sum_{i}^{m} \rho(r_i) \tag{6.13}$$

Where $\rho(r_i)$ is a symmetric positive definite function with a unique minimum at zero, z = measurement vector, x is the state vector, and $h(x)$ is the measurement function. There are a number of M-estimators that have been suggested by various researchers. M-estimators seek the vector $\hat{x}$ that minimizes the cost function given by (6.13). Some of these estimators are described here. In these estimators the tuning parameter α used in the following functions is to be specified by the user and it can affect $J(x)$ significantly. If the selected value of α is small, it may result in $J(x)$ with multiple local extrema. If the value of α is taken to be large to minimize this problem it may result in reducing the effect of bad data suppression. Normally, the value of α is selected to lie between 1 and 4.

When quadratic constant criterion is used we have

$$\rho(r_i) = \frac{r_i^2}{\sigma_i^2} \text{ if } \left|\frac{r_i}{\sigma_i}\right| \le \alpha \tag{6.14}$$

and

$$\rho(r_i) = \frac{r_i^2}{\sigma_i^2} \text{ if } \left|\frac{r_i}{\sigma_i}\right| > \alpha \tag{6.15}$$

If a quadratic linear criterion is used then we have

$$\rho(r_i) = \frac{r_i^2}{\sigma_i^2} \text{ if } \left|\frac{r_i}{\sigma_i}\right| \le \alpha \tag{6.16}$$

and

$$\rho(r_i) = 2\alpha\sigma_i |r_i| - \alpha^2\sigma_i^2 \text{ if } \left|\frac{r_i}{\sigma_i}\right| > \alpha \tag{6.17}$$

If a square root criterion is used, then

$$\rho(r_i) = \frac{r_i^2}{\sigma_i^2} \text{ if } \left|\frac{r_i}{\sigma_i}\right| \le \alpha \tag{6.18}$$

$$\rho(r_i) = 4.\alpha^{3/2}\sqrt{r_i/\sigma_i} - 3\alpha^2 \text{ if } \left|\frac{r_i}{\sigma_i}\right| > \alpha \tag{6.19}$$

Various objective functions are shown in Figure 6.2(a), (b), and (c).

6.4 State Estimation Methods with Bad Data Rejection Properties

There have been numerous attempts to develop one-step algorithms that simultaneously reject bad data and estimate the state of a power system [8]. These methods that are based on nonquadratic cost function become a least square estimator in the absence of bad data. The basic idea here is to change the cost function so that the effect of bad data is minimized. Some of these methods are described below.

6.4.1 Methods Using Nonquadratic Objective Functions

Application of a nonquadratic objective function to a state estimation is used for bad data rejection. The conventional quadratic objective function of WLS state estimator is converted to a quadratic constant objective function as shown in Figure 6.2(a). The Lagrangian multiplier method for optimization can be used for these problems. In the Lagrange multiplier method one additional variable is introduced for each constraint. Thus if the original problem has n variables and m equality constraints, then by adding m additional variables, the final number of unknowns become $m + n$. For example, if the problem is having two variables and one constraint as: minimize $\mathcal{L}$ subject to λ, the necessary condition using Lagrange function can be written as

$$\mathcal{L}(x_1, x_2, \lambda) = f(x_1, x_2) + \lambda g(x_1, x_2) \tag{6.20}$$

The first-order necessary conditions for minimizing (6.13) can be derived using the Lagrangian equation as

$$\mathcal{L} = \sum_{i=1}^{m} \rho(r_i) + \lambda(z - h(x) - r) \tag{6.21}$$

whereas the necessary conditions for minimum of $j(r)$ are

$$\frac{\partial L}{\partial \lambda} = z - h(x) - r = 0 \tag{6.22}$$

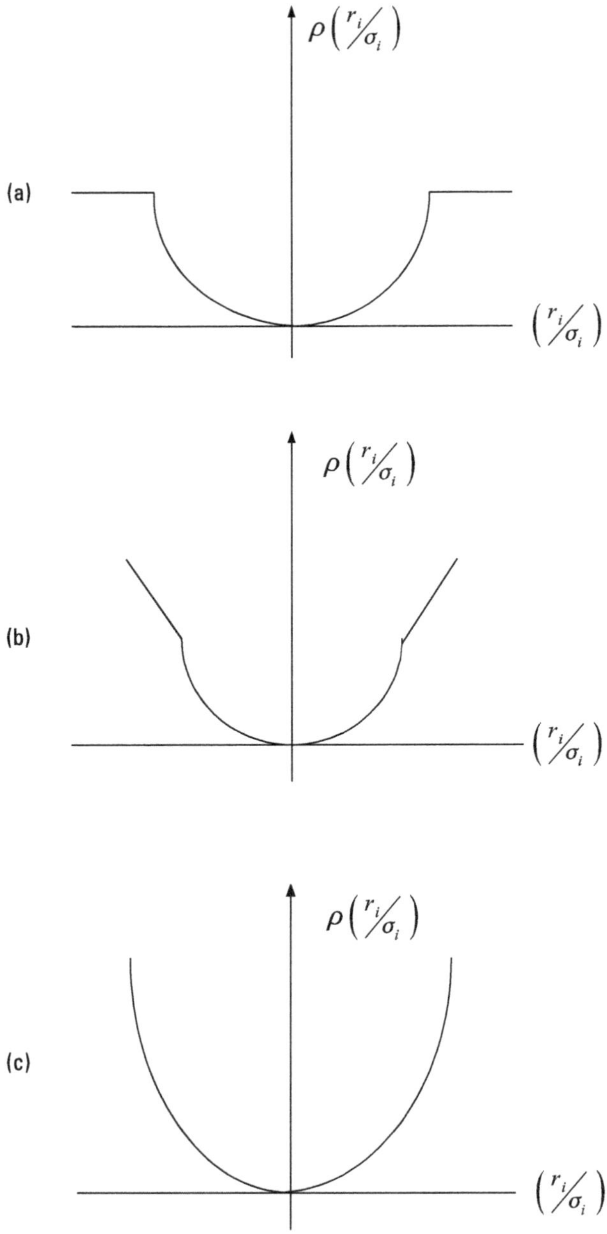

Figure 6.2 (a) Quadratic constant objective function, (b) quadratic linear objective function, and (c) square root objective function.

$$\frac{\partial L}{\partial r} = \frac{\partial \rho_r}{\partial r} - \lambda = 0 \tag{6.23}$$

$$\frac{\partial L}{\partial x} = -H^T \lambda = 0 \tag{6.24}$$

where

$$H = \frac{\partial h(x)}{\partial x}$$

Eliminating λ from (6.23) and (6.24), the reduced set of necessary conditions are

$$\frac{\partial L}{\partial \lambda} = z - h(x) - r = 0 \tag{6.25}$$

$$\frac{\partial L}{\partial x} = -H^T \lambda = H^T \frac{\partial \rho_r}{\partial r} = 0 \tag{6.26}$$

These equations can be solved iteratively using the Gauss-Newton method. Using the first-order Taylor's approximation for linearization:

$$h(x^{(k+1)}) \approx h(x^k) + H(x^k)\Delta x^k \tag{6.27}$$

If $\frac{\partial \rho_r}{\partial r} = \Upsilon$,

$$\Upsilon\left(r^{k+1)}\right) \approx \Upsilon\left(r^k\right) + \nabla\Upsilon\left(\mathrm{r}^{\mathrm{k}}\right)\Delta r \tag{6.28}$$

where $\Delta x^k = x - x^k$,

$$\nabla\Upsilon = \partial\Upsilon\left(r^k\right)/\partial\left(r^k\right) \tag{6.29}$$

$$r^k = z - h\left(x^k\right) \tag{6.30}$$

The necessary conditions are rewritten as a set of iterative linear system equations that can be solved for Δx and r.

$$r^{(k)} = Hx^{(k)}\Delta x + r \tag{6.31}$$

and
$$H^T \Upsilon(r^{k)} = H^T \left[\nabla\Upsilon\right]^{-1} H \Delta x^{(k)} \tag{6.32}$$

Now, rewriting the Gauss-Newton equations in terms of Δx and r:

$$\begin{bmatrix} \nabla\Upsilon & \bar{H} \\ \bar{H}^T & 0 \end{bmatrix} \begin{bmatrix} r \\ \Delta x^{(k)} \end{bmatrix} = \begin{bmatrix} \nabla\Upsilon \mathrm{r}^{(\mathrm{k})} \\ \bar{H}^T r^{(k)} - H^T.\Upsilon r^{(k)} \end{bmatrix} \tag{6.33}$$

where

$$H = \nabla\Upsilon \cdot Hx^k \tag{6.34}$$

This symmetric linear system is solvable provided $\bar{H}$ has a full column rank. Measurements whose $\nabla\Upsilon = 0$ are set aside from calculation because their residuals exceed threshold value. In such a case the system will be rank-deficient and will be solved only for those terms whose $\nabla\Upsilon \neq 0$.

6.5 Least Absolute Value State Estimators

Techniques based on the least absolute value (LAV) minimization a derivative of linear programming have shown potential for use in state estimators. These estimators are useful in detecting and eliminating bad data from the measurement set. The basic difference between the least square (LS) and LAV estimators is that a best LAV approximation is obtained by minimizing the sum of the absolute values of the residuals, whereas a best LS approximation is obtained by minimizing the sum of the squares of the residuals [6, 9].

The solution to any LAV problem can be found using one of the many LAV algorithms available, most of which use some form of linear programming. The LAV estimator can be considered as a nonquadratic estimator in which the range of the quadratic objective function is reduced to zero and a linear objective is used as shown in Figure 6.3.

The LAV or WLAV state estimation problem is stated as

$$\text{minimize } j(x) = \sum_{i=1}^{m} w_i \left|r_i\right| \tag{6.35}$$

Subject to

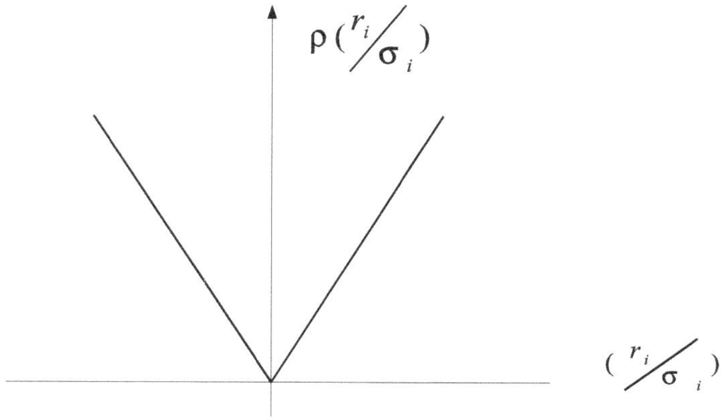

Figure 6.3 Least absolute value objective function.

$$z = h(x) + r \tag{6.36}$$

or

$$j(x) = \sum_{i=1}^{m} \frac{|z_i - h(x)_i|}{\sigma_i} \tag{6.37}$$

Here w is the m-dimensional vector of weights. The weights are based on the measurement of standard deviations. A higher weight signifies a more reliable measurement. The solution to this minimization problem is given by n basic measurements, which fits perfectly the state estimate and have zero residuals. The remaining n-m measurements are nonbasic variables and can have nonzero residuals.

6.6 Simplex Method

The WLAV problems are shown above as linear programming problems; hence these can be solved using the well-known simplex method. Equation (6.38) can be linearized using the Taylor series expansion and taking the first term only as

$$z = z_0 + \Delta z = h(x_0) + H(x_0) \cdot \Delta x + r \tag{6.38}$$

The state estimates can be approximately found by iterative estimation of state correction vector Δx by the equation

$$\Delta x = H(x_0) \cdot \Delta x + r \tag{6.39}$$

if $r = \eta - \rho$ where η and ρ are vectors of deviation variables.

In estimating Δx, the objective function will be the weighted sum of deviation variables. This function can be minimized by introducing the slack variables in the measurement equations and setting up the cost function as

$$\min J = \sum_{i=1}^{m} w_i (\eta_i + \rho_i) \tag{6.40}$$

where w_i is the weight of each measurement. A larger weight signifies a more reliable measurement. The constraints are the equations relating the state vector to the measurements.

$$z = h(x) + \eta - \rho \tag{6.41}$$

The constraints are to be linearized at every step.

$$h(x^k) = h(x^{k-1}) + H(x^{(k-1)}) \Delta x^k \tag{6.42}$$

Or

$$H(x^{(k-1)}) \Delta x^k + \eta^k - \rho^k = z - h(x^{k-1}) = \Delta z^k \tag{6.43}$$

This equation represents the linearized constraints at a given starting point. The simplex method is used to find Δx and the current starting point is then updated. The constraints are again linearized at the updated operating point and the procedure is repeated until convergence is obtained. The m components of η provide a starting basis. Out of these n are replaced by the n components Δx during the simplex procedure. The m-n components of η and ρ that remain in the basis at nonzero values correspond to the bad data provided there is sufficient redundancy in measurements. If the number of bad data is more than m-n, then this method will not perform satisfactorily.

The choice of starting point is very important. The starting point should be not far from the solution; otherwise, the number of linearizations required for convergence will be large. In a real-time implementation, the results of the previous estimation can be used as a starting point.

6.7 Interior Point Algorithm

The interior point method for linear programming was introduced by Karmakar in 1984 [11]. For many problems it is competitive or superior to simplex method (see Appendix 6A for a description of linear programming). Many versions of interior point methods have been developed and represent significant development in the theory and practice of linear programming. From a theoretical point of view they represent algorithms that use interesting geometric ideas. From a practical point of view they provide solution to large-scale problems that arise in many applications. There are a number of interior point methods available for the solution of WLAV problems. Two categories of methods that work satisfactorily are affine scaling variants and barrier function methods [12, 13].

The WLAV state estimation problem is of the form

$$\min \sum_{i=1}^{m} |r_i| \tag{6.44}$$

subject to

$$z = h(x) + r \tag{6.45}$$

and

$$\eta \cdot \rho \geq 0 \tag{6.46}$$

where $r = \eta - \rho$, η and ρ represent the positive and negative parts of r, respectively.

The logarithmic barrier function was first used by Frisch and consists of applying Newton's method to the following family of problems parameterized by the barrier parameter $\mu > 0$.

$$\min(\eta + \rho) - \mu \sum_{i=1}^{m} (l_n \eta_i + l_n \rho_i) \tag{6.47}$$

The solution to these problems for a sequence of decreasing approach the solution of original problem. Various strategies for updating μ are available in literature. The variables Δx are unrestricted and do not figure in the barrier function. The Newton search direction is given by the Karush-Kuhn-Tucker first-order optimality conditions for the quadratic approximation of (6.47) subject to retaining feasibility.

The Lagarangian $\mathcal{L}$ can be written as

$$\mathcal{L}=\sum_{i=1}^{m}\left[\eta_i+\rho_i-\mu l_n\eta_i-\mu l_n\rho_i-\lambda_i\left(z_i-h_i(x)-\eta_i+\rho_i\right)\right] \quad (6.48)$$

Applying Karush-Kuhn-Tucker condition for minima

$$\mathcal{L}_x=\frac{\partial\mathcal{L}}{\partial x}=\sum_{i=1}^{m}\lambda_i\frac{\partial h_i(x)}{\partial x}=0 \quad (6.49)$$

$$\mathcal{L}_{\lambda_i}=\frac{\partial\mathcal{L}}{\partial\lambda_i}=z_i-h_i(x)-\eta_i+\rho_i=0 \quad (6.50)$$

$$\mathcal{L}_{\eta_i}=\frac{\partial\mathcal{L}}{\partial\eta_i}=1-\mu\left(\eta_i\right)^{-1}-\lambda_i=0 \quad (6.51)$$

$$\mathcal{L}_{\rho_i}=\frac{\partial\mathcal{L}}{\partial\rho_i}=1-\mu\left(\rho_i\right)^{-1}+\lambda_i=0 \quad (6.52)$$

Representing the variables ρ, $\eta\lambda$, and h_i by their first-order approximations and $H=\frac{\partial h}{\partial x}$, $h^T(x)=[h_1(x), h_2(x) \dots h_m(x)]$, $\lambda^T=[\lambda_1, \lambda_2, \dots \lambda_m]$. Equation (6.49) can be written as

$$\mathcal{L}_x=H^T\lambda_i^0+H^T\Delta\lambda=0 \quad (6.53)$$

$$\mathcal{L}_{\lambda_i}=\frac{\partial\mathcal{L}}{\partial\lambda_i}=z_i-h_i(x)-\eta_i+\rho_i=0 \quad (6.54)$$

6.8 LMS Estimator

The LMS estimator is a high breakdown state estimator [13]. This is a generalization of the idea that the median of a set of real values is a more robust estimate than the mean. In the unidimensional and simple regression case, the LMS estimator minimizes not the sum of the squared residuals but the median, and hence is called the LMS estimator. The LMS estimator for $n =$ 1 or 2 is

$$\min j(x) = \left(\frac{r_i}{\sigma_i} \right)^2 \tag{6.55}$$

This means that for all alternative state estimates x, it selects the one having a minimum median squared residual.

Since a nonzero residual occurs only when there are redundant measurements (i.e., $m > n$), then for multidimensional states $n > 2$ the median is given by

$$md = \frac{m}{2} + \frac{n+1}{2} \tag{6.56}$$

where $m =$ number of measurements and $n =$ number of states.

In the multiple regression case it minimizes

$$j(x) = \left(\frac{r_{md}}{\rho_{md}} \right)^2 \tag{6.57}$$

Here $|x|$ denotes the integer part of the real number x. In order to determine the sample of N nonredundant measurements and to determine the minimum surplus (or redundancy level) of the available measurement sets, a number of techniques are available in the literature. Basically, the samples with n measurements, each for which the network is minimally observable are drawn, and for each one of these, states are calculated. Weighted residuals of the remaining m-n measurements are computed. The optimum solution is given by the sample that minimizes the objective function of (6.49).

The computation of LMS estimates is not straightforward. This is a difficulty common to all high breakpoint estimators. The procedures used are extension of combinatorial search techniques applied to nonlinear models.

In principle, the combinatorial search should include $\begin{bmatrix} m \\ n \end{bmatrix}$ samples, which leads to very large computational time even for small systems. However, it is possible to select only a certain number k of random selections such that the probability P for at least one sample is not contaminated is nearly 1. The relationship between P and k can be found from the following. If ε denotes the assumed fraction of gross errors among the m measurements and m is large, then ε may be considered as the probability of drawing just one bad data. Now the probability of selecting n good data is $(1-\in)^n$. Then the probability of selecting k samples of bad data of size n is $\left[1-(1-\in)^n\right]^k$.

Then

$$P = 1 - \left[1 - (1-\in)^n\right]^k \tag{6.58}$$

In a power system the zero injections or pseudomeasurements must be incorporated in any drawn sample. As such, their number must be subtracted from the number n of the unknowns when computing k. It is desirable to use a fast decoupled LMS estimator wherever it is possible.

References

[1] Adibi, M. M., and D. K. Thorne, "Remote Measurement Calibration," *IEEE Transactions on Power Systems,* Vol. 1, 1986, pp. 194–203.

[2] Adibi, M. M., and J. P. Stoval, "On Estimation of Uncertainties in Analog Measurements," *IEEE Transactions on Power Systems,* Vol. 6, 1991, pp. 1222–1230.

[3] Hodges, J. L., Jr., "Efficiency in Normal Samples and Tolerance of Extreme Values for Some Estimates of Location," *Proceedings of the Fifth Berkley Symposium on Mathematical Statistcs and Probability, Volume 1: Statistics,* Berkley, CA: University of California Press, 1967, pp.163–168.

[4] Hampel, F. R., "A General Qualitative Definition of Robustness," *Annals of Mathematical* Statistics, Vol. 42, 1971, pp. 187–1896.

[5] Donoho, D. L., and P. J. Huber, *The Notion of Breakdown Point,* Belmont, CA: Wadsworth Statist./Probab. Ser., 1983, pp. 157–184.

[6] Celik, M. K., and A. Abur, "A Robust WLAV State Estimation Using Transformations," *IEEE Transactions on Power Systems,* Vol. 7, No. 1, 1992, pp. 106–114.

[7] Rousseeuw, P. J., and B. C. van Zomeren, "Unmasking Multivariate Outliers and Leverage Points," *Journal of the American Statistical Association,* Vol. 85, 1990, pp. 633–639.

[8] Irving, M. R., R. C. Owen, and M. J. H. Sterling, "Power System State Estimation Using Linear Programming," *Proceedings of the IEE,* Vol. 125, 1978, pp. 879–885.

[9] Kotiuga, W. W., and M. Vidyasagar, "Bad Data Rejection Properties of Weighted Least Absolute Techniques Applied to Static State Estimation," *IEEE Transactions on Power Apparatus and Systems,* Vol. PAS-101, No. 4, 1982, pp. 844–853.

[10] Chen, J.,and A. Abur, "Placement of PMUs to Enable Bad Data Detection in State Estimation," *IEEE Transactions on Power Systems,* Vol. 21, No. 4, 2006, pp. 1608–1615.

[11] Karmarkar, N. "A New Polynomial Time Algorithm for Linear Programming," *Combinatorica,* Vol. 4, 1984, pp. 373–395.

[12] Zhuang, F., and R Balasubramanian, "Bad Data Suppression in Power System State Estimation with Variable Quadratic Constant Criterion," *IEEE Transactions on Power Apparatus and Systems,* Vol. 104, No. 4, 1985, pp. 857–863.

[13] Nammi, M., J. B. A. London, Jr., and A. C. B. Delbem, et al., "Robust State Estimator Based on Least Median of Squares Method," *Transmission and Distribution Conference and Exposition Latin America,* August 13–15, 2008, pp. 1–6.

Appendix 6A: Linear Programming

Linear programming is an optimization method for problems in which the objective function and the constraints are linear functions of decision variables. The constraint equations in linear programming can be in the form of equality or inequality equations.

The general linear programming problem can be written in the standard form as

$$\min f\left(x_1, x_2 \ldots x_n\right) = \sum_{j=1}^{n} c_j x_j \tag{6A.1}$$

Subject to constraints

$$\sum_{j=1}^{n} a_{ij} x_j = b_i,\ i = 1, 2, \ldots, m \tag{6A.2}$$

$$x_j \geq 0, j = 1, 2, \ldots, n \tag{6A.3}$$

It can be seen that there are m equations in decision variables (unknown) in a linear programming problem. For optimization the condition is that $m < n$. This case corresponds to an underdetermined set of linear equations, which, if they have one solution, will have an infinite number of solutions.

The linear programming is used to find the solution that minimizes the function f. For $m = n$ there is no possibility of optimization as it will give an exact solution. For > n, there are $m - n$ redundant equations that can be eliminated. The simplex method is a powerful scheme for obtaining the optimal solution.

6A.1 Simplex Algorithm

The first step in the simplex method is to construct an auxiliary problem by adding certain artificial variables into the original linear programming problem. The main purpose of adding artificial variables is to convert the auxiliary problem into a canonical form from which its basic solution can be obtained. From the basic solution we check to see whether it is optimum. If it is not optimum then another feasible solution is obtained. This process is continued until the optimum solution is obtained. For details about linear programming solution using the simplex method, any book on optimization techniques may be consulted, such as *Engineering Optimization: Theory and Practice* by Rao.

7

State Estimation Using Line Current Measurements

7.1 Introduction

In Chapter 3, state estimation techniques were described that accept the voltage magnitudes, real and reactive power of bus injections, and line flow measurements to determine the state of the system. Most of the state estimation techniques are also developed for transmission systems of the power system. As well, an increasing interest to automate at the subtransmission and distribution levels has resulted in a requirement for state estimation in these networks. However, for medium-voltage networks the amount of real-time information is very limited, which cannot provide a satisfactory solution to a state estimation problem. At the subtransmission level, the basic measurements are the branch currents and voltage magnitudes at the nodes only. As well, current magnitude measurements are often made on transformers connecting loads to the transmission system. Since state estimation is only possible if there are enough measurements distributed in such a manner that the system is observable, the insufficient number of measurements available at the subtransmission and distribution levels may not guarantee the observability of a complete system The current measurement systems provide the information about magnitude only, and no information is available about the phase angle. Therefore, when parts of a network contain only branch current measurements, the unique solution of state estimation can be obtained only if additional information in one of the following forms is available:

1. Voltage magnitude and direction of active power injection $P_i \leq 0$, or $P_i \geq 0$;
2. Active power flow or injection measurement and the direction of reactive power injections;
3. The reactive power measurements and direction of active power injections $P_i \leq 0$, or $P_i \geq 0$.

The power system state estimation problem can be solved uniquely when the measurement set contains voltage magnitude; real and reactive power flow, and injection measurements only. However, when branch current magnitude measurements are also considered, then there may be several possible solutions for the same measurement set. In such cases, the estimator may converge to any one of these possible solutions depending on the initialization values. This problem will be discussed later. However, one clear advantage of voltage and current measurements is the simplicity with which prefiltering can be achieved. Voltage magnitudes differing more than a certain threshold from 1 p.u. can be easily filtered. Similarly, currents at both line ends must be almost equal, and the readings can be discarded if the difference is large. Sometimes the square of the current is used instead of the current.

7.2 Modeling State Equations

In order to incorporate current magnitude measurements in a state estimation program, an equation for the magnitude of current measurement in terms of state variables (V, θ) must be obtained. If the normal equivalent representation as shown in Figure 7.1 is used, the line current measurement between nodes i and j can be written as follows:

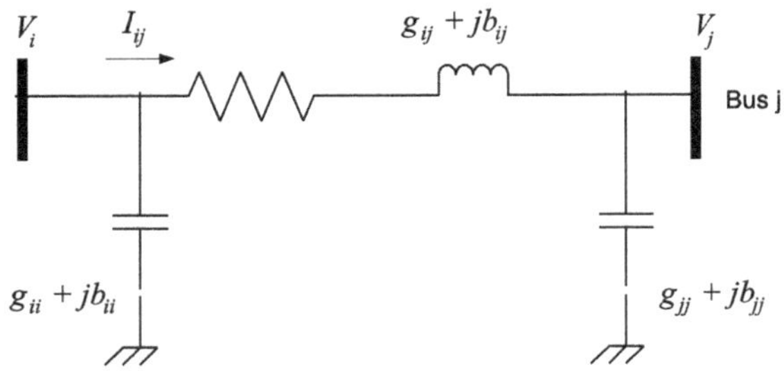

Figure 7.1 π model of line.

$$I_{ij} = \left[AV_i^2 + BV_j^2 - 2CV_iV_j\cos(\theta_i - \theta_j]^{1/2} \right] \tag{7.1}$$

where

$$\begin{aligned} A &= \left(g_{ij} + g_{ii}\right)^2 + \left(b_{ij} + b_{ii}\right)^2 \\ B &= g_{ij}{}^2 + b_{ij}{}^2 \\ C &= g_{ij}\left(g_{ij} + g_{ii}\right) + b_{ij}\left(b_{ij} + b_{ii}\right) \end{aligned} \tag{7.2}$$

where g_{ij}, b_{ij}, g_{ii}, and b_{ii} are the π equivalent parameters of the transmission line, as shown in Figure 7.1.

The partial derivatives of the current magnitude measurements with the state variable can be written as

$$\frac{\partial |I_{ij}|}{\partial V_i} = \frac{1}{I_{ij}}\left[AV_i - CV_j\cos(\theta_i - \theta_j)\right] \tag{7.3}$$

$$\frac{\partial |I_{ij}|}{\partial \theta_i} = \frac{1}{I_{ij}}\left[CV_iV_j\sin(\theta_i - \theta_j)\right] \tag{7.4}$$

$$\frac{\partial |I_{ij}|}{\partial V_j} = \frac{1}{I_{ij}}\left[BV_j - CV_i\cos(\theta_i - \theta_j)\right] \tag{7.5}$$

$$\frac{\partial |I_{ij}|}{\partial \theta_j} = \frac{1}{I_{ij}}\left[-CV_iV_j\sin(\theta_i - \theta_j)\right] \tag{7.6}$$

In the state estimation problem, these partial derivatives will be the entries in the Jacobian matrix. The most common algorithm used in state estimates is WLS method described earlier and repeated here.

$$\min\ j(x) = [z - h(x)]^T R^{-1}\left[z - h(x)\right] \tag{7.7}$$

The first-order optimality condition leads to

$$\frac{dj}{dx} = H^T(x)R^{-1}\left[z - h(x)\right] = 0 \tag{7.8}$$

The most common method of solving (7.8) is by using the iteration method. The problem with this method of solution is that with current measurements, the iterations cannot be started from a flat voltage profile. That is, for $V_i = V_j = 1.0$ p.u., and $\theta_i = \theta_j = 0.0°$.

$$\left|I_{ij}\right| = (g_{ii}^2 + b_{ii}^2)^{1/2} \cong b_{ii} \tag{7.9}$$

If b_{ii} is also small, the current becomes very small during a flat start condition. Then the solution to (7.8) may not be possible. Such a problem can be minimized to a great extent by taking the square of rms current value, I_{ij}^2. The equation for the square of the rms current is

$$\left[I_{ij}\right]^2 = AV_i^2 + BV_j^2 - 2CV_iV_j\cos(\theta_i - \theta_j] \tag{7.10}$$

The Jacobian matrix entries are

$$\frac{\partial\left|I_{ij}\right|^2}{\partial V_i} = 2[AV_i - CV_j\cos\left(\theta_i - \theta_j\right) \tag{7.11}$$

$$\frac{\partial\left|I_{ij}\right|^2}{\partial \theta_i} = 2[CV_iV_j\sin\left(\theta_i - \theta_j\right) \tag{7.12}$$

$$\frac{\partial\left|I_{ij}\right|^2}{\partial V_j} = 2[BV_j - CV_i\cos\left(\theta_i - \theta_j\right) \tag{7.13}$$

$$\frac{\partial\left|I_{ij}\right|^2}{\partial \theta_j} = 2[-CV_iV_j\sin\left(\theta_i - \theta_j\right) \tag{7.14}$$

At flat start conditions, (7.11) to (7.14) can be evaluated, provided ampere measurements are not critical. The problem of small Jacobian terms can be minimized in estimators run in tracking mode. In these estimators instead of a flat start, the iterative solution starts with the values of previously estimated states.

7.3 State Estimation with Current Measurements

The line current measurements were not considered for use in state estimation due to the fact that there is a possibility of multiple solutions if only the current and voltage measurements are used [4]. Also mathematical problems arise if a flat start is used in the iterative solutions, as shown in Section 7.2. When there are lines with little line charging susceptance and magnitude current flows along them are measured, some of the entries in the Jacobian measurements become undefined. Another difficulty is related to network observability. Since more than one likely solution is possible for the state estimation problem with current measurements, the system observability is not guaranteed by the state estimation solution.

However, many utilities want to incorporate line current measurements in state estimation solutions so that ampere RTU/transducer errors may be detected. Also, those parts of the system that are not properly measured can be made observable with line current measurements. However, it must be made clear that a current cannot be used as a linking measurement on its own to provide observability. It needs a complimentary measurement with it to make the system observable. The current measurements therefore are not considered for the purpose of providing observability to the system. However, if it is important to use current measurement for observability, I_{ij} may also be substituted by power measurement P_{ij} [6]. If observability is guaranteed by other types of measurements, the line current measurements can be included for improving redundancy.

7.3.1 Multiple Solutions

If only voltage and current measurements are available without any power measurements, the information about phase angle can be obtained from (7.1).

$$\cos\left(\theta_i - \theta_j\right) = \frac{AV_i^2 + BV_j^2 - I_{ij}^2}{2CV_iV_j} \tag{7.15}$$

The solution to this equation results in two values of $(\theta_i - \theta_j)$. This will result in two different power flow solutions, as P_{ij} and I_{ij} are functions of $(\theta_i - \theta_j)$ as can be seen from the following equations.

$$P_{ij} = V_i^2\left(g_{ii} + g_{ij}\right) - V_iV_j\left[g_{ij}\cos(\theta_i - \theta_j) + b_{ij}\sin(\theta_i - \theta_j)\right] \quad (7.16)$$

$$Q_{ij} = V_i^2\left(b_{ii} + b_{ij}\right) - V_iV_j\left[g_{ii}sin(\theta_i - \theta_j) - b_{ij}\cos(\theta_i - \theta_j)\right] \quad (7.17)$$

In conventional state estimation the problem of multiple solutions is not an issue, as it is assumed that the measurements are available in pairs in the form of $\delta - Q$ or $P - V$.

7.4 Methods to Obtain a Unique Solution in the Presence of Current Measurements

The problem of ambiguity in the real power flow direction posed by the measurement set containing a large number of current measurements and its solution using different methods is discussed in this section. The first method proposed in [2] is state estimation with nonlinear equality constraints. The basic procedure proposed in this section is logical extension of the state estimator with the equality constraints discussed in Chapter 3. The basic idea here is to use other information, besides real-time measurements that does not need to be measured, to discard many solutions that may be produced. For example, in subtransmission networks, active power is always drawn from buses connected to the distribution side via transformers. Few buses of subtransmission receive active power drawn from the transmission level or small generators. It is therefore possible to know with certainty whether specified real power for each node is less, greater than, or equal to zero. Mathematically this information can be incorporated in the form of inequality constraints. The problem can be expressed in the form of (7.7) subject to the inequality constraints

$$c_i(x) \geq 0$$

and

$$c_i\left(x\right) = \pm\sum_{i=1}^{N} V_iV_j\left\{\cos\left(\theta_i - \theta_j\right) + b_{ij}\sin\left(\theta_i - \theta_j\right)\right\} \quad (7.18)$$

Simple Methods Using Kirchhoff's Law for a Unique Solution

Assuming that voltage magnitude in all the buses are known, then it is possible to find (p, θ) by applying Kirchhoff's node or loop equation. In a closed loop, if either current or power flow is measured in every branch, then all branches will be observable provided at least one branch has a power flow measurement. If only current is measured in all branches of the loop, then there will be two possible solutions corresponding to two values of angle ($\pm(\theta_i - \theta_j)$ for each branch.

Applying Kirchhoff's law for any given injection measured node, if all the branches connected to that node are measured either in the form of current or power then all branches will be observable.

7.5 Determining the Uniqueness of a Solution Based on Numerical Methods

For simple systems the heuristic methods as well as topological methods are available to find the uniqueness of a solution. However, for large systems that are coupled, numerical techniques can be used. The critical current measurements leading to multiple solutions can be detected with the help of entries in the residual covariance matrix. The covariance matrix of residuals can be obtained as follows: using the WLS state estimation, the measurement residuals are related to measurement errors as

$$r = We \tag{7.19}$$

where

$$W = (I - H\left(H'R^{-1}H\right)^{-1} H^T R^{-1} \tag{7.20}$$

The covariance matrix of residuals is

$$C = WR = R - H\left(H'R^{(-1)}H\right)^{(-1)} H^T \tag{7.21}$$

If a measurement is critical the corresponding column in the C matrix will be zero. Also, if the measurement z_j belongs to residual spread component, the jth column of the C matrix will also be zero, except for the entries corresponding to the measurements belonging to same residual spread component.

As we are interested in only determining the critical measurements for the current magnitude measurements, the columns corresponding only to current measurements must be checked. The entry to the column j of matrix C can be calculated as follows.

Determine y_j from

$$H'R^{-1}Hy_j = h'_j \tag{7.22}$$

where h_j is the jth row of H.

The jth column of C is

$$C_j = R_j - Hy_j \tag{7.23}$$

where R_j is the jth column of R containing all zeros except R_{jj}.

To find the multiplicity of a solution, it is assumed that the system is solvable but may have more than one solution. Compute the columns of C corresponding to current measurements and check for a nonzero entry corresponding to power flow or an injection measurement. If it is nonzero skip that column. Otherwise there is a possibility of multiple solutions corresponding to that current measurement. If the column has zero entry then the current magnitude measurement is critical. If the system has a critical measurement, the columns of C should be checked for threshold value and accordingly whether the system is observable will be decided.

7.6 Bad Data Detection in the Presence of Current Measurements

If the measurement data also contains the current measurement, the identification of bad data is affected by their presence. As discussed in Section 7.1, in the presence of line current measurements, the system having n state variables is not uniquely observable with n linearly independent measurements only. It is therefore important to classify the measurements as critical and noncritical. Critical measurements containing bad data cannot be detected by simple state estimation algorithms.

7.6.1 WLS State Estimation

In least square estimation the bad data detection method is based on a normalized residual test and covariance matrix and is carried out after the state estimation. As discussed in Chapter 5, the measurement residuals are calculated after obtaining the estimated value. Then the diagonal elements of residual covariance matrix are obtained. The normalized measurement residuals are obtained by dividing the measurement residuals with the standard deviation σ. The normalized residual are sorted in descending order. Now the largest normalized residual is compared against the threshold value. If it is less than the threshold then the measurement having the largest normalized residual is eliminated as it is considered to be bad data. The same procedure is repeated until all the bad measurements are eliminated.

When only conventional measurements are present, the possibility of eliminating a critical measurement is almost zero. The reason for this is that the critical measurements in conventional measurements have zero normalized residuals; hence there is no chance of identifying them as bad measurements. However, when current measurements are also present the above condition is not valid. It is therefore essential to check at the end of each cycle that the measurement being deleted is noncritical. The condition for noncriticality is tested applying the following rule. First we define the *residual spread component*. When a redundant (noncritical) measurement is in error, the set of measurements that will be affected by this measurement forms the residual spread component corresponding to these measurements.

- If the measurement with the largest normalized residual does not belong to any of the residual spread components containing at least one current measurement, then that measurement can be removed without any problem (noncritical) [9].
- For other cases the measurements must be checked for criticality of a unique solution. If a measurement is a power flow or injection measurement and the residual spread component does not contain any other power flow or injection measurement, then solution may not be unique. Hence, this measurement cannot be eliminated straightaway unless checked as given below.
- If any of the current measurements in the same residual spread component will become critical if this measurement is removed, then this measurement cannot be discarded. Otherwise, it can be removed.

7.6.2 WLAV Estimation

WLAV estimation, as discussed in Chapter 5, is fundamentally different from the WLS method. It satisfies a subset of m measurements exactly while the remaining n-m measurements have nonzero residuals. Another difference is that WLAV methods are used to identify the bad measurements without correcting or reestimating the states. If the measurement set contains only the conventional measurements, the estimated state is optimal, since the bad measurements are already left out by the WLAV process. But when current measurements are also present, the WLAV estimator results have to be tested further to check for optimality. The normalized residuals of suspected bad data must be tested for criticality by the three conditions described above. If any of the data is suspected to be uniquely critical; that is, it will have the possibility of multiple solutions, then a WLAV solution is not acceptable.

7.7 Some Observations

Earlier research in the inclusion of current measurements in state estimation has concluded that current measurements are useful in improving the observability of the network. Once the system is observable, inclusion of current measurements is beneficial in improving the accuracy of estimated states up to a certain saturation level.

In the area of bad data detection, there are only certain cases where the addition of the current measurements is helpful in detecting the bad data. However, the main beneficial effect of line current measurements is to raise the detection probability in power injection measurements.

Thus the addition of current measurements in state estimation is advisable if they are available in a system that with conventional measurements has observability but does not have the desired degree of redundancy. However, with the availability of phasor measurement units (PMU), which provide the phasor current values, the application of current measurements in state estimation will provide improvements in all areas of state estimation. The application of PMUs in state estimation is discussed in Chapter 8.

References

[1] Handschein, E., M. Langer, and E. Kliokys, "An Interior Point Method for State Estimation with Current Magnitude Measurements and Inequality Constraints," *IEEE Proceedings of Power Industry Computer Application Conference,* May 7–12, 1995, Salt Lake City, UT, pp. 385–391.

[2] Ruiz Munoz, J. M., and A. G. Exposito, "A Line Current Measurement Based State Estimator," *IEEE Transactions on Power Systems,* Vol. 7, No. 2, 1992, pp. 513–518.

[3] Geisler, K. I., "Ampere Magnitude Line Measurements for Power System State Estimation," *IEEE Transactions on Power Systems,* Vol. PAS 103, No. 8, 1984, pp. 1962–1964.

[4] Abur, A., and A. G. Exposito, "Detecting Multiple Solutions in State Estimation in the Presence of Current Magnitude Measurements," *IEEE Transactions on Power Systems,* Vol. 12, No. 1,1997, pp. 370–375.

[5] Marinho, J. L., P. A. Machado, and C. Bongers, "On the Use of Line Current Measurements for Reliable State Estimation in Power Systems," *Power Industry Computer Applications Conference,* PICA-79, Cleveland, OH, May 15–19, 1979, pp. 2–6.

[6] Abur, A., and A. G. Exposito, "Multiple Solutions and Unique Observability in State Estimation," *Power System Computation Conference,* Dresden, Germany, 1996, pp 1200–1206.

[7] Abur, A., and A. G. Exposito, "Observability and Bad Data Identification When Using Ampere Measurements in State Estimation," *IEEE International Conference on Circuits and Systems,* Chicago, 1993.

[8] Abur, A., and A. G. Exposito, "Bad Data Identification When Using Ampere Measurements," *IEEE Transactions on Power Systems,* Vol. 12, No. 2, 1997, pp. 831–836.

[9] Exposito, A. G., and A. Abur, "Generalized Observability Analysis and Measurement Classification," *IEEE Transactions on Power Systems,* Vol. 13, No. 3, 1998, pp. 1090–1095.

8

Phasor Measurement Units Applications in State Estimation

8.1 Synchronized Phasor Measurements

The real-time monitoring of energy system makes use of state estimation extensively. This is traditionally accomplished by the state estimator that resides in the control center computer and has access to the measurements received from numerous substations in the monitored system. These measurements are commonly provided by the remote terminal units (RTU) at the substations and include real/reactive power flows, power injections, and magnitudes of bus voltages and branch currents. Providing the accurate measurements and complete topology of the system to the control center computer is not currently possible, and data update and communication is slow and can take several seconds. Currently state estimators use the Inter-Control Center Communications Protocol (ICCP) gathering asynchronously fed data with an arrival rate of one sample per 4 to 10 seconds. On the other hand, time-synchronized devices like PMU are capable of directly measuring the positive sequence of the voltage phasor for a given system. Synchronized phasor measurement units were first introduced in the 1980s, and have since become a mature technology with many applications including in state estimation of power system. A PMU as per the definition (IEEE 1344, IEEE C37.118) is a device (mostly microprocessor-based) that reports the magnitude and phase angle of an analog and/or derived phasor with respect to the global time reference and is shown in Figure 8.1(a).

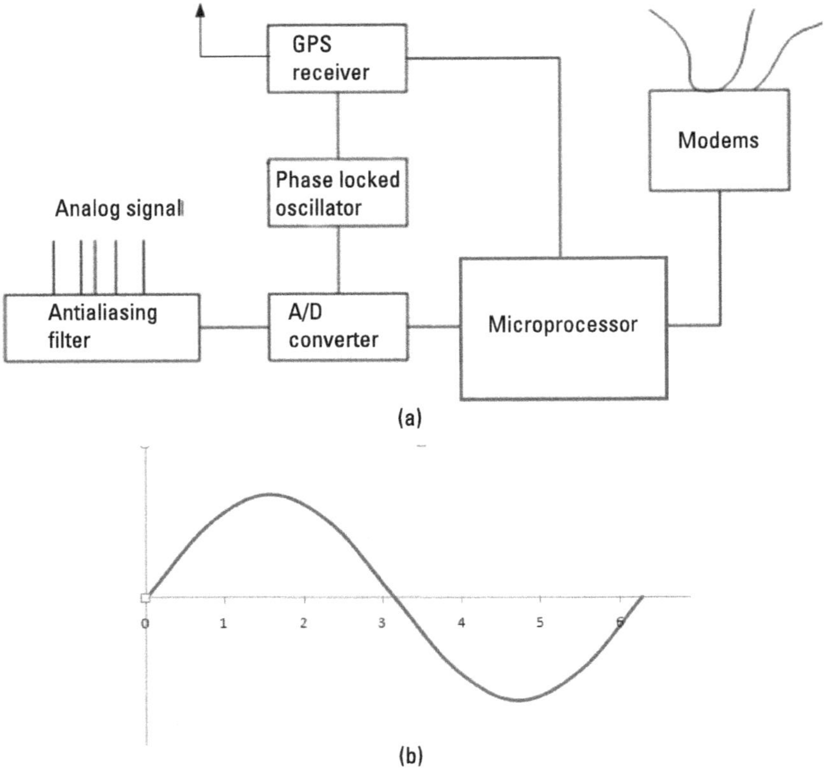

Figure 8.1 (a) Block diagram of PMU and (b) sinusoidal signal and its phasor representation.

A PMU can measure voltage and current phasors at the rate of 20 to 60 times in a second. It is, therefore, possible to measure the state of the system directly, provided we have enough PMUs in the system. While it is not possible to have PMUs in every bus of the system because of their cost, an optimum number of these devices can improve the state estimation in robustness and accuracy.

8.1.1 Synchrophasor Definition

Synchrophasor is a term used to describe a phasor that has been estimated at an instant known as the time tag of the synchrophasor. In order to obtain simultaneous measurements of phasor across a wide area of the power system, it is necessary to synchronize these time tags. Phasor representation of a sinusoidal signal is given by

$$x(t) = X_m \cos(\omega t + \phi) \tag{8.1}$$

$$x(t) = \frac{X_m}{\sqrt{2}} e^{j\phi} = \frac{X_m}{\sqrt{2}} (\cos\phi + j\sin\phi) \tag{8.2}$$

As can be seen from (8.2), frequency ω is not expressed in phasor representation. The sinusoidal signal and its phasor representation are shown in Figure 8.1(b).

Since the frequency of the phasor is implicit in the phasor diagram, all phasors included in a single phasor diagram must have the same frequency. Phasors are usually introduced as a means of representing steady state sinusoidal waveforms of fundamental power frequency. However, in a power system when it is not quite in steady state, phasors can still be used to describe its behavior as in most of the cases the variations in waveforms of voltage and current are slow. In a digital measuring system, samples of the waveform for one nominal period are collected starting at $t = 0$. The fundamental component of the waveform is obtained using discrete Fourier transform as

$$X = \frac{\sqrt{2}}{N} \sum_{k=1}^{N} x_k e^{-jk2\pi/N} \tag{8.3}$$

where N is the total number of samples in one period and x_k are the N samples of the waveform. A PMU provides the phasor using the sampled data of the input signal. The phasor X as obtained from N samples is the correct representation of the fundamental component of the waveform when other transient components are present. It is essential that the input waveform be filtered before applying to PMU to eliminate aliasing error in DFT calculation.

Phasors can be measured for each of the three phases, and the positive sequence phasor can be obtained from the relation as

$$X_1 = \frac{1}{3}\left(X_a + \alpha X_b + \alpha^2 X_c\right) \text{ where } \alpha = e^{j2\pi/3} \tag{8.4}$$

8.2 Application of PMUs in Static State Estimation

The first PMU using the Global Positioning Systems (GPS) transmissions to synchronize the sampling clocks was developed at the Virginia Polytechnic Institute by Arun Phadke in 1988. The first commercial PMU was produced in 1991 by the Macrodyne company in the United States. The advantages of GPS are quite visible in such diverse areas as car navigation systems and power systems. Today, in fact, car drivers as well as most taxi drivers around the world are using GPS. PMUs provide real-time synchronized phasor measurements in a power system with better than 1 μs synchronization accuracy that is obtained from GPS [1, 2]. The PMU measurements can provide real-time snapshots of the states of monitored buses in a power system. Thus, apart from standard real and reactive power and voltage and current measurements available from conventional meters, the PMUs can also provide voltage and current phasor measurements (Figure 8.2).

Simplicity of relations between the phasor measurements and state variables means that state estimation using phasor measurements is much quicker than classical state estimation. Another feature of the PMU measurements is that it is not necessary to have a completely observable network before state estimation could be performed. Since the cost of PMU is quite high, it is unlikely that a SCADA system will be completely replaced by PMUs in near future. It is, therefore, useful to develop methods of state estimation that will use the input from both a SCADA system and PMUs. In this case, it is possible to combine all the measurements in one set and develop a hybrid nonlinear state estimator. In recent work, an alternative approach has been suggested where it is possible to retain the traditional state estimator software intact, and add the phasor measurements in a linear post-processing step. In this chapter, several new developments in state estimation with phasor

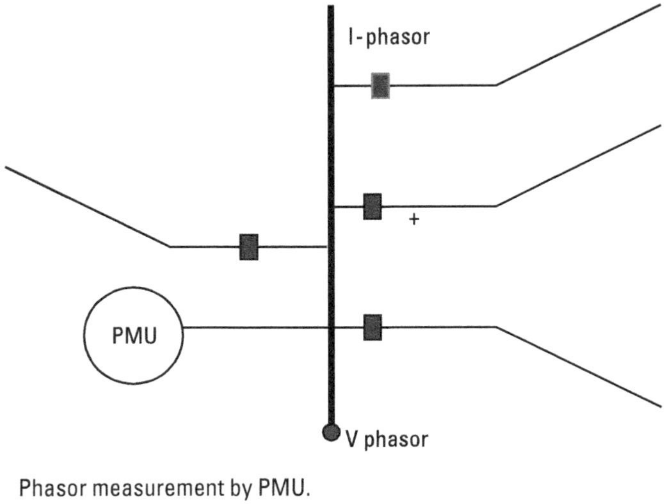

Figure 8.2 Phasor measurement by PMU.

measurements such as bad data processing and topology error processing are discussed.

8.2.1 Linear Measurement Model with PMUs

A PMU installed at a specific bus is capable of measuring not only the bus voltage phasor, but also the current phasors along all the lines incident to that bus. Thus in addition to the phasor voltage at this bus, it is possible to compute the phasor voltages of all its neighboring buses, making them observable. All the observable buses can be divided into two levels: the *directly observable buses* with PMUs installed on them and *indirect observable buses* with PMUs installed in nearby buses. If measurements provided by PMUs alone are considered, then a measurement model can be obtained as follows.

8.2.2 Measurement Equations

The measurement equation for directly observable buses can be written as

$$z_i = V_i + ev_i \tag{8.5}$$

where z_i is the measured complex voltage (V_i, θ_i) at bus i, V_i is the "true" complex voltage at bus i, and ev_i is the complex measurement noise of this PMU. Thus for N buses in the system, if there are M directly observable buses, the measurement equation in the matrix form can be written as

$$z_M = [I][V_{\mathrm{LM}}] + v_{\mathrm{V}} \tag{8.6}$$

where z_M is the complex voltage measurement subvector, I is the identity matrix, V_{LM} is the true complex voltages at all directly observable buses, and v_V is the voltage measurement noise subvector.

8.2.3 Indirectly Observable Buses

If the bus i is indirect observable, then for each PMU placed at some adjacent bus the measurement equation can be obtained using current measurement phasors. For the ith indirectly observable bus, there is at least one PMU placed at the adjacent bus j. Then for each PMU placed at some adjacent bus j, the measurement equation will be

$$z_{ji} = \begin{bmatrix} Y_{ji} & -Y_{ji} \end{bmatrix} \begin{bmatrix} V_j \\ V_i \end{bmatrix} + ev_j \tag{8.7}$$

where z_{ji} is the complex measured current at bus j (towards bus i), Y_{ji} is the admittance of line (j,i), and are the "true" complex voltages at bus j and i, respectively, and ev_j is the complex measurement noise of the PMU at bus j.

Thus for complete N bus system with M buses having PMUs, the measurement equation can be written as

$$z_C = \begin{bmatrix} Y_{ML1} & Y_{ML2} \end{bmatrix} \begin{bmatrix} V_{L1} \\ V_{L2} \end{bmatrix} + e_{vc} \tag{8.8}$$

where z_C is the complete complex current measurement vector, Y_{ML1}, Y_{ML2} are the line admittance matrices that relates, respectively, the directly and indirectly observable bus voltages to the current measurement vector. V_{L1} and V_{L2} are true complex voltages at directly and indirectly observable buses, respectively, and e_{vc} is the current measurement noise vector. In this equation the measurement matrix $[Y_{ML1}\ Y_{ML2}]$ has each row sum up to zero. If the number of PMUs installed is sufficient, then there may be more than one bus having PMU near the indirectly observable bus. In such cases the contribution of all the buses having PMU must be included in the nearby bus not having PMU.

Combining the measurements of direct and indirect observable buses, the measurement equation can be written as

$$\begin{bmatrix} z_V \\ z_C \end{bmatrix} = \begin{bmatrix} I & 0 \\ Y_{ML1} & Y_{ML2} \end{bmatrix} \begin{bmatrix} V_{L1} \\ V_{L2} \end{bmatrix} + \begin{bmatrix} ev_V \\ e_{vc} \end{bmatrix} \tag{8.9}$$

The above equation in the simple form can be written as

$$z' = H''V + ev' \tag{8.10}$$

8.3 Combining Conventional and PMU Measurements

An important aspect of state estimation is the speed at which it processes data coming from measurement devices. Conventional measurements are provided by the SCADA system, which is designed to capture only quasi-state operating conditions. In actual operation the snapshot is collected over a few seconds. The refresh rate of PMU measurements can be up to 50 or 60 times in a second, and each measurement from a PMU contains a time stamp. The time information that the PMU extracts from the GPS signals can be used to find out the exact time at which the measurements were taken. This can help in eliminating the situation where due to the communication delays the measurement data coming from the system are retrieved at different times. Generally, the advantage of taking data from PMUs in addition to conventional measurement devices is that it makes the data more accurate, but it also transforms the existing critical measurements into the redundancy measurements. It is very important from a bad data detection point of view as bad data cannot be detected from critical measurements.

Direct state measurement could be accomplished if PMUs are employed at all system buses. If branch currents are also measured they would provide enough redundancy in the measurement set. A phasor-based state estimator requires PMUs located at all buses that are desired to be monitored by SCADA/EMS system. However, the presently most utilities are installing PMUs gradually. It is, therefore, inevitable that state estimation will be carried out with PMUs and conventional measurements together.

There are various procedures for combining the phasor measurements with traditional P-Q measurements in state estimation. Basically these methods can be divided into two categories. In the first category are those methods that combine the traditional measurements and the phasor measurements in a hybrid state estimator, as shown in Figure 8.3.

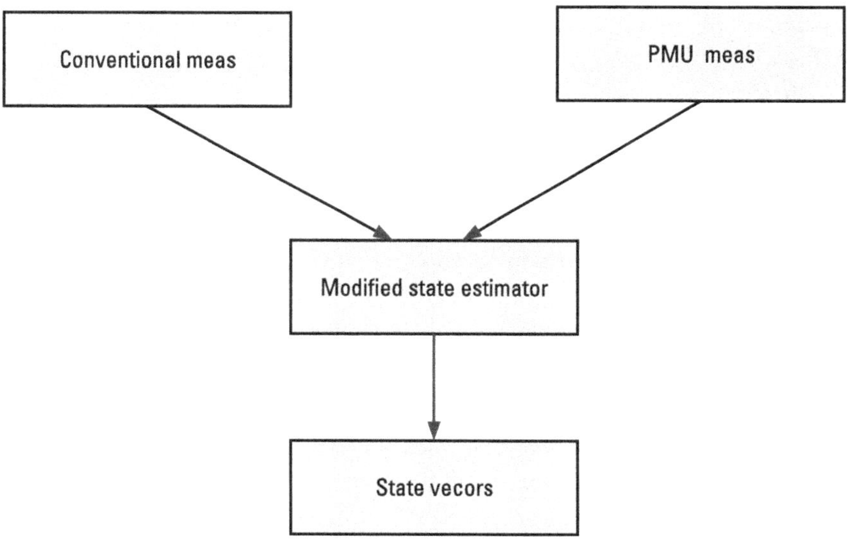

Figure 8.3 Hybrid state estimator.

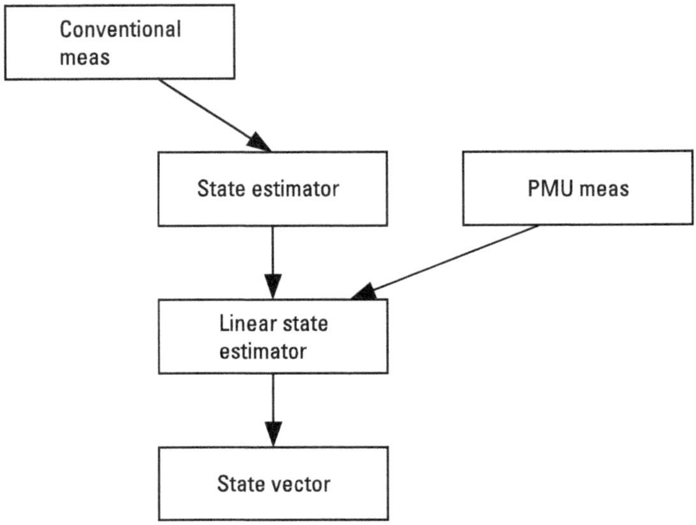

Figure 8.4 Hybrid state estimator with PMU measurement.

Due to the different nature of complex phasor measurements, the direct inclusion of phasor measurements in state estimators requires significant modification to the state estimation software. The other more attractive option is to achieve the state estimates from the traditional estimator and then incorporate the phasor measurements in a linear estimator, as shown in Figure 8.4.

In the first method, both the voltage and current phasor measurements are included in the traditional WLS state estimator. In this model, the voltage phasor measurements are used in the polar coordinates denoted as magnitude V and angle, which directly correspond to magnitude and angle of the bus where it is measured. The current measurements by PMUs can be directly incorporated in a state estimator by any one of three possible ways:

1. By applying current phasor magnitude and phase angle;
2. Applying a real and imaginary part of the complex current measurement;
3. Applying a pseudovoltage measurement using the current phasor measurement and known line parameters.

The asynchronous time stamp of conventional measurements can be combined with the synchronous time stamp of PMU. In case there is no common instant corresponding to measurements, one set of measurements can be interpolated. All the phase angles measured and estimated are referred to a common reference that is the phase angle of the slack bus. It is assumed here that a PMU is installed at the slack bus.

8.3.1 Formulation of the Hybrid State Estimator

For the conventional measurements the nonlinear state estimation model as discussed in Chapter 3 and is repeated here

$$z = h(x) + e \tag{8.11}$$

The estimate of the state of the system using the WLS method is obtained by determining the state vector that minimizes the function:

$$J(x) = [z - h(x)]^T R^{-1}[z - h(x) \tag{8.12}$$

This estimate should satisfy the first-order optimality constraint as

$$g(x) = \frac{\partial J(x)}{\partial x} = -H^T R^{-1}\left[z - h(x)\right] = 0 \tag{8.13}$$

The iterative equations using the Gauss-Newton method is as follows:

$$x^{k+1} = x^k + (H^T R^{-1} H^{-1})^{-1} H^T R^{-1}(z - h(x^k) \tag{8.14}$$

where z is the measurement vector consisting of conventional as well as PMU measurements, and H is the Jacobian matrix of mixed measurements.

Consider the first case where the phasor magnitude and phase angle of the current measured by the PMUs are directly included in the existing state estimator. The modified Jacobian matrix with measurements from PMU included will be

$$
[H]=\begin{bmatrix}
\dfrac{\partial P_{inj}}{\partial \theta} & \dfrac{\partial P_{inj}}{\partial V} \\
\dfrac{\partial P_{flow}}{\partial \theta} & \dfrac{\partial P_{flow}}{\partial V} \\
\dfrac{\partial Q_{inj}}{\partial \theta} & \dfrac{\partial Q_{inj}}{\partial V} \\
\dfrac{\partial Q_{flow}}{\partial \theta} & \dfrac{\partial Q_{flow}}{\partial V} \\
\dfrac{\partial V_{meas}}{\partial \theta} & \dfrac{\partial V_{meas}}{\partial V} \\
\dfrac{\partial \theta_{V,meas}}{\partial \theta} & \dfrac{\partial \theta_{V,meas}}{\partial V} \\
\dfrac{\partial \theta_{I,meas}}{\partial \theta} & \dfrac{\partial \theta_{I,meas}}{\partial V} \\
\dfrac{\partial I_{meas}}{\partial \theta} & \dfrac{\partial I_{meas}}{\partial V}
\end{bmatrix} \tag{8.15}
$$

Here P_{inj} and Q_{inj} represent the real and reactive power injection measurements, and P_{flow} and Q_{flow} are the real and reactive power flow measurements, and θ_V and θ_I are the phase angles of the bus voltages and the line currents, respectively, measured by the PMUs, and V_{meas} and I_{meas} are the voltage and current magnitudes recorded by the PMUs.

Assuming a pi-model of the transmission line as shown in Figure 8.5, the elements of the submatrices of conventional measurements of the

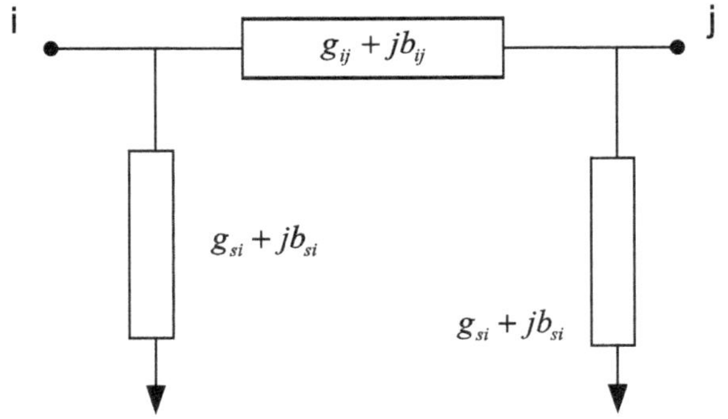

Figure 8.5 Pi-model of a transmission line.

Jacobian matrix $\frac{\partial P_{inj}}{\partial \theta}, \frac{\partial P_{inj}}{\partial V}$ and so forth can be determined by differentiating the power flow equation.

The submatrices $\frac{\partial \theta_{V,meas}}{\partial \theta}$ and $\frac{\partial V_{meas}}{\partial V}$ are identity matrices, and the terms $\frac{\partial \theta_{V,meas}}{\partial V}, \frac{\partial V_{meas}}{\partial \theta}$ are zero.

The current phasor I_{ij} in line *i-j* is related to bus voltages V_i and V_j by the equation given below.

$$\begin{aligned} I_{ij} &= V_i\left(g_{si} + jb_{si}\right) + (V_i - V_j)(g_{ij} + jb_{ij}) \\ &= V_i \cos\theta_i \left(g_{si} + g_{ij}\right) - V_i \sin\theta_i \left(b_{si} + b_{ij}\right) \\ &\quad + b_{ij} V_j \sin\theta_j - g_{ij} V_j \cos\theta_j + V_i \cos\theta_i \left(b_{si} + b_{ij}\right) \\ &\quad + V_i \sin\theta_i \left(g_{si} + g_{ij}\right) - b_{ij} V_j \cos\theta_j - g_{ij} V_j \sin\theta_j \\ &= C + jD \end{aligned} \tag{8.16}$$

The magnitude and phase angle of current I_{ij} is given by

$$\left|I_{ij}\right| = \sqrt{C^2 + D^2}, \ \theta_{ij} = \tan^{-1}\frac{D}{C} \tag{8.17}$$

Here $(g_{si} + jb_{si})$ is the shunt admittance connected at bus *i*, and $(g_{ij} + jb_{ij})$ is the series impedance of the line between buses *i* and *j*. The partial derivatives $\frac{\partial \theta_{I,meas}}{\partial V}$ can be obtained by differentiating (8.17).

If we assume that the measurements are independent of each other, the diagonal elements of the error covariance matrix will have only diagonal elements except for phase angles. The diagonal elements for measurements other than the phase angles are determined by the standard uncertainties in the measurements as

$$R(i,i) = u_i^2 \tag{8.18}$$

where u_i is the uncertainty in the *i*th measurement.

8.3.2 Applying Real and Imaginary Parts of the Complex Current Measurement

In the second method, the current phasor measured by the PMU is first decomposed into real and imaginary parts before including it into the state estimation program. The real and imaginary parts of currents flowing from bus *i* to *j* measured by PMU is obtained from (8.16) as

$$I_{ij} = C + jD \tag{8.19}$$

The Jacobian matrix will be modified as given below in (8.20).

The error covariance matrix is also modified using the computed values of $u_{I_{ijreal}}$ and $u_{i_{ijimag}}$ for the corresponding elements. Here, $u_{I_{ij}}$ and $u_{\theta_{ij}}$ are the standard uncertainties associated with current magnitude and phase angle measured by PMU.

8.3.3 Applying Pseudovoltage Measurements

$$[H] = \begin{bmatrix} \dfrac{\partial P_{inj}}{\partial \theta} & \dfrac{\partial P_{inj}}{\partial V} \\ \dfrac{\partial P_{flow}}{\partial \theta} & \dfrac{\partial P_{flow}}{\partial V} \\ \dfrac{\partial Q_{inj}}{\partial \theta} & \dfrac{\partial Q_{inj}}{\partial V} \\ \dfrac{\partial Q_{flow}}{\partial \theta} & \dfrac{\partial Q_{flow}}{\partial V} \\ \dfrac{\partial V_{meas}}{\partial \theta} & \dfrac{\partial V_{meas}}{\partial V} \\ \dfrac{\partial \theta_{V,meas}}{\partial \theta} & \dfrac{\partial \theta_{V,meas}}{\partial V} \\ \dfrac{\partial I_{real}}{\partial \theta} & \dfrac{\partial I_{real}}{\partial V} \\ \dfrac{\partial I_{imag}}{\partial \theta} & \dfrac{\partial I_{imag}}{\partial V} \end{bmatrix} \tag{8.20}$$

In the third method, the voltage phasor at a bus connected to a bus having PMU is obtained by using the voltage and current phasor measurements at the PMU bus and the known parameters of the transmission line. For pi representation of the transmission line of Figure 8.5 assuming that the PMU is connected at bus *i*, the voltage phasor at bus *j* is given by

$$V_j = \frac{V_i\left(g_{ij} + jb_{ij}\right)\left(+g_{si} + jb_{si}\right) - I_{ij}}{\left(g_{ij} + jb_{ij}\right)} = V_{jreal} + V_{jimag} \tag{8.21}$$

The Jacobian matrix is modified to have partial derivatives of the voltage phase angle and magnitude of both the direct and the pseudomeasurements as given in (8.22).

The off-diagonal nonzero elements of pseudomeasurements can be obtained by differentiating (8.21) after expanding it in rectangular coordinates.

The standard uncertainty in the measurement of pseudovoltage magnitude and phase angle at bus *j*, obtained by using the PMU measurement at bus *i* can be obtained as in (8.23).

$$[H] = \begin{bmatrix} \frac{\partial P_{inj}}{\partial \theta} & \frac{\partial P_{inj}}{\partial V} \\ \frac{\partial P_{flow}}{\partial \theta} & \frac{\partial P_{flow}}{\partial V} \\ \frac{\partial Q_{inj}}{\partial \theta} & \frac{\partial Q_{inj}}{\partial V} \\ \frac{\partial Q_{flow}}{\partial \theta} & \frac{\partial Q_{flow}}{\partial V} \\ \frac{\partial \theta_V}{\partial \theta} & \frac{\partial \theta_V}{\partial V} \\ \frac{\partial V}{\partial \theta} & \frac{\partial V}{\partial V} \end{bmatrix} \tag{8.22}$$

$$u_{V_{ij}} = \sqrt{\sum_{k=1}^{4} \left(\frac{\partial V_j}{\partial p_k} \right)^2 u_{pk}^2} \tag{8.23}$$

If there are any changes in the measurement, the hybrid state estimators will be modified accordingly. For addition of any PMU or removal from the measurement system, the rows in the Jacobian matrix corresponding to current and voltage measurements by that PMU can be added or removed accordingly.

8.3.4 Post-Processing Adding of Phasor Measurements

In formulating the linear state estimation problem incorporating phasor measurements as a post-processing step, the state vector and the measurement vector are expressed in rectangular coordinates. Now the new measurement set will consist of the output of traditional state estimator in rectangular coordinates combined with the voltage and current measurements of PMU. The phase angles of all the buses are measured with respect to the angle of the swing bus having zero angle. The error covariance matrix is modified to reflect the transformation of state variables to rectangular form.

8.4 Phasor Measurements in Dynamic State Estimation

In earlier chapters, the state estimation of the power system is assumed to be in steady state when the measurements are made. Under normal conditions the power system generally remains in a quasistatic state. Hence in order to have continuous monitoring of a power system, state estimation must be performed at short intervals of time. However, in large power systems the static state estimation fails to capture the dynamic nature of the power system. This led to the development of dynamic state estimation (DSE) algorithms that calculate the state of the system by modeling the time-varying behavior of the power system. The DSE algorithms can predict the state of the system one step ahead, which has tremendous advantages in performing security analysis. This proves to be a major advantage for the operator as it gives more time to take control decisions in case of emergencies.

The accuracy of the state estimation depends on the measurements used for estimation. Dynamic state estimation is similar to static state estimation except that the network states can be predicted from previous network state values. It can be shown that the static state estimation step is numerically equivalent to the update step for a dynamic state estimator where a priori network state estimate is included as additional measurements for static state estimator. As a first step, a mathematical model of the system including PMU for dynamic state estimation is obtained. To have simplification in the model only PMU-based voltage magnitude and phase angles are used and not the current measurements. As already discussed in this chapter, current measurements cannot be directly incorporated in the WLS state estimator.

In PMU-based DSE, decoupled WLS can be used as active and reactive measurements can be dealt separately. Measurements from PMU will be used with higher weights compared to weights of SCADA measurements. The PMU-based dynamic state estimation algorithm is as follows:

- The state vector is initialized at *k*th instant of time with a flat start ($V = 1.0$).
- Obtain the measurement vector with higher weights assigned to PMU-based measurements than SCADA measurements.
- Perform the initial PMU included static state estimation to obtain the states at the instant *k*.
- Using the exponential smoothing technique, obtain the predicted state vector for instant *k*.
- Get the new measurement set containing both normal and PMU measurements for $k+1$ instant.
- Calculate the measurement residual from the new measurements at $k+1$ and predicted measurements calculated from predicted state vector.
- The next step in DSE is the state filtering step for which extended Kalman filtering may be used.
- Once the estimation process is complete, the angle and voltage estimates corresponding to PMU measurements are replaced with the PMU measurements themselves.

8.5 Optimal PMU Placement

Since the cost of PMUs and its communication network is large, it is desirable to employ an optimum number of PMUs so that all buses in the system are observable. The optimum PMU placement problem basically requires the determination of minimum number of PMUs required along with their placement locations so that the system is completely observable. There are two approaches for optimum placement:

1. Using graph-theoretic observability approach;
2. Integer programming based numerical approach.

The graph theoretic approach or topological approach is more suitable for systems that have a large number of conventional measurements. For such systems, which may be unobservable or have few observable islands, the topological method will find the strategic locations of PMUs by merging the observable islands. Since only boundary buses will contribute to the merging of observable islands, these will be the strategic locations for placing PMUs.

The placement of a minimal set of PMUs so as to make the system measurement model observable and thereby linear is described here. A PMU placed at a bus measures the voltage as well as all the current phasors at that bus, requiring the extension of the topological observability theory. In particular, the concept of a spanning tree is extended to that of a spanning measurement subgraph with an actual or a pseudomeasurement assigned to each of its branches. The minimal PMU set is found through a dual-search algorithm that uses both a modified bisecting search and a simulated-annealing-based method. The former fixes the number of PMUs while the latter looks for a placement set that leads to an observable network for a fixed number of PMUs. In order to accelerate the procedure, an initial PMU placement is provided by a graph-theoretic procedure that builds a spanning measurement subgraph according to a depth-first search.

Spanning Tree Approach

This algorithm uses a spanning tree method to obtain a solution for complete observability of the system. In this method, the concept of depth of observability is used. The level of unobservability is defined as the number of directly linked unobservable buses connected to an observed bus which does not have a PMU installed at that bus.

In the spanning tree method, a root node of a network is chosen by the user. It then branches off to the terminal nodes at the end of the branch. The terminal nodes are of two types: a parent terminal node that is at the end of main branch; and a spanning tree terminal node that is an intermediate branch from the path. After reaching the terminal branch, it again returns to the splitting point of the branch and explores other directions. After terminal nodes are reached, it will return to the root node. The first PMU is placed at the first bus after the root node. The next PMU is placed according to the following rule

$$d_p = u + 3$$

where d_p is the distance in terms of number of buses from the current PMU to the next PMU to be placed and is the desired level of unobservability. The flow chart of the spanning tree approach is shown in Figure 8.6.

When a PMU is placed one bus away from the spanning tree bus so as to avoid placing a PMU at the radial bus, it then backtracks to the previous split and searches this area until the entire network is searched and the root node is reached. To determine the minimum number of PMUs, the same procedure is repeated with different root nodes.

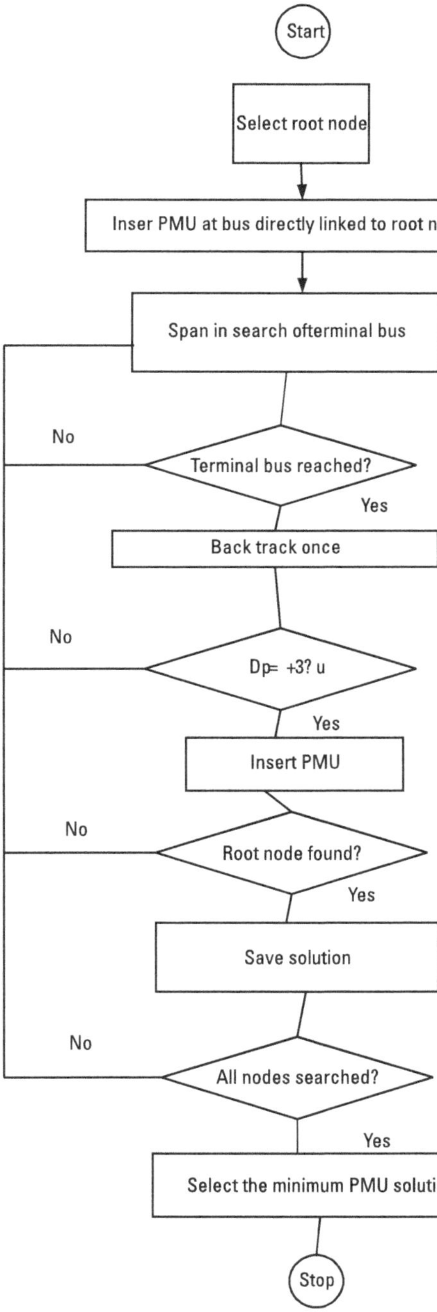

Figure 8.6 Flow chart for the spanning tree approach algorithm.

8.5.1 Integer Programming-Based Optimal Placement

In deciding the optimal number of PMUs an assumption is made that each PMU has enough number of channels to measure bus voltage and all incident branch current phasors at a given bus. For an n bus power system the optimal PMU placement problem can be stated as

$$\min \sum_{i}^{n} w_i x_i \tag{8.24}$$

Subject to

$$f(X) \geq \hat{1}$$

where n is number of buses in the system, w_i is the cost of installing the PMU at bus i. $f(x)$ is a vector function representing constraints and its rows correspond to one or more observable islands corresponding to each critical measurement. Vector X is a binary (0/1) decision variable whose entries are defined as

$$x_i = \begin{cases} 1 & \text{if a PMU is installed at bus } i \\ 0 & \text{if no PMU is installed} \end{cases} \tag{8.25}$$

1 is a vector whose entries are all 1. The constraint equations will be obtained for three different conditions based on locations and types of measurements:

- Without conventional measurements;
- PMU measurements and injections;
- PMU measurements, flow measurements, and injections.

For a system having no conventional measurements and injections, the PMU placement problem may be considered as finding the minimum of PMUs such that a bus must be *reached* at least once by the set of PMUs. To begin with, binary connectivity matrix A is formed. The elements of matrix A are defined as

$$A_{i,j} = \begin{bmatrix} 1 \text{ if } i = j \\ 1 \text{ if } i \text{ and } j \text{ are connected} \\ 0 \text{ for all other cases} \end{bmatrix} \tag{8.26}$$

$$\begin{aligned} &f(X) \geq 1 \\ &X = [x_1, x_2, \dots x_N]^T \\ &x_i \in \{0,1\} \end{aligned} \tag{8.27}$$

PMU and Injection Measurements

When injection measurements are available that also include zero injections, the constraints are modified so that the requirement of number of PMUs is reduced. It is seen that one constraint can be removed by applying Kirchhoff's current law at the zero injection bus, as will be shown by Example 8.1.

PMU, Injections, and Flow Measurements

In a case where all three types of measurements are available, the number of constraints will be modified and eliminated. Here three cases are demonstrated by an example for a simple nine-bus system as shown in Figure 8.7. For complex systems having a large number of buses, TOMLAB's optimization toolbox can be used.

Example 8.1

Find the optimal number of PMUs for the nine-bus sample system shown in Figure 8.7.

Solution

No conventional measurements

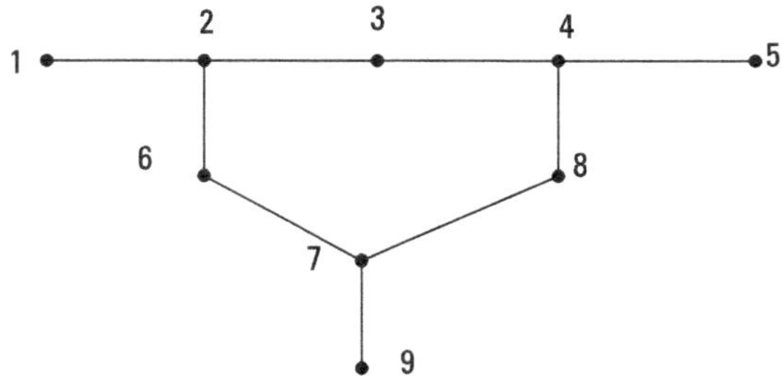

Figure 8.7 A nine-bus sample system.

$$\min x_1 + x_2 + x_3 + \ldots + x_9 \tag{8.28}$$

Subject to the constraints AX > = 1

Matrix A can be directly obtained from the bus admittance matrix by transforming its entries into binary form. For a nine-bus system the matrix A is

$$A = \begin{bmatrix} 1 & 1 & 0 & 0 & 0 & 0 & 0 & 0 & 0 \\ 1 & 1 & 1 & 0 & 0 & 1 & 0 & 0 & 0 \\ 0 & 1 & 1 & 1 & 0 & 0 & 0 & 0 & 0 \\ 0 & 0 & 1 & 1 & 1 & 0 & 0 & 1 & 0 \\ 0 & 0 & 0 & 1 & 1 & 0 & 0 & 0 & 0 \\ 0 & 1 & 0 & 0 & 0 & 1 & 1 & 0 & 0 \\ 0 & 0 & 0 & 0 & 0 & 1 & 1 & 1 & 1 \\ 0 & 0 & 0 & 1 & 0 & 0 & 1 & 1 & 0 \\ 0 & 0 & 0 & 0 & 0 & 0 & 1 & 0 & 1 \end{bmatrix} \tag{8.29}$$

$$X = [x_1 x_2 \ldots x_9]^T \tag{8.30}$$

The constraints therefore, can be written as

$$\begin{aligned} f_1 &= x_1 + x_2 \geq 1 \\ f_2 &= x_1 + x_2 + x_3 + x_6 \geq 1 \\ f_3 &= x_2 + x_3 + x_4 \geq 1 \\ f_4 &= x_3 + x_4 + x_5 \geq 1 \\ f_5 &= x_4 + x_5 \geq 1 \\ f_6 &= x_2 + x_6 + x_7 \geq 1 \\ f_7 &= x_6 + x_7 + x_8 + x_9 \geq 1 \\ f_8 &= x_4 + x_7 + x_8 \geq 1 \\ f_9 &= x_7 + x_9 \geq 1 \end{aligned} \tag{8.31}$$

If we put all the constraint as equal to 1, it means that for bus 1 the PMU must be placed either at bus 1 or bus 2 to make it observable. Similarly for bus 2, the PMU may be placed at1, 2, 3, or 6 to make it observable. For a simple system like this we can find the minimum number of PMUs by inspection. For example in this case if a PMU is placed at bus 2, buses 1, 2, 3, and 6 are observable. By placing PMU at bus 7, buses 6, 7, 8, and 9 become observable. For buses 4 and 5 the PMU can be placed either at bus 4 or 5. Thus by placing PMUs at buses 2, 4, and 7, all the buses will be observable. If linear integer programming optimization is used, the following four solutions are obtained: (1, 4, 7), (2, 4, 7), (2, 5, 7) and (2, 4, 9).

The same problem is solved for the system having injections including zero injections. Here bus 2 is assumed to have zero injections. For zero injection bus the net injected current is zero. It is, therefore, clear that if the phasor voltages at any three of the four buses 1, 2, 3, and 6 are known, then the fourth bus voltage can be obtained by using Kirchhoff's current law. The constraints of these buses therefore are modified as

$$f_1 = x_1 + x_2 + f_2 . f_3 . f_6 \geq 1$$
$$f_3 = x_2 + x_3 + x_4 + f_1 . f_2 . f_6 \geq 1$$
$$f_6 = x_2 + x_6 + x_7 + f_1 . f_2 . f_3 \geq 1$$

This means that bus 1 is observable if a PMU is placed in either bus 1 or 2 or the phasor voltages are known in buses 2, 3, and 6. If the value of is substituted in

$$f_1 = x_1 + x_2 + (x_1 + x_2 + x_3 + x_6) . f_3 . f_6 \geq 1$$

If A and B are two subsets, where A is subset of B, then A · B = A, and A + B = B using this simplification the expression for

$$f_1 = x_1 + x_2 + x_3 + x_4 + x_6 \geq 1$$
$$f_3 = x_2 + x_3 + x_4 + x_1 + x_6 \geq 1$$
$$f_6 = x_2 + x_6 + x_7 + x_1 + x_4 \geq 1$$

all other constraints remain unchanged. The constraint for bus 2 is not required.

If all three types of measurements are present the constraint equations will be modified as follows: If a flow measurement between branch *i* and *j* is available, the bus voltage phasor of *i* bus can be obtained if voltage phasor at bus *j* is known (or the *j*th bus voltage can be found if the *i*th bus voltage is known). If flow measurement between bus 1 and 2 is available the constraints of bus 1 and 2 are merged into a joint constraint.

$$f_{1\,merged} = f_1 + f_2$$

Or

$$f_{1\,merged} = x_1 + x_2 + x_3 + x_6 \geq 1$$

Loss of a Single PMU

Although PMUs are devices that have very high reliability, there is still a chance that one out of many PMUs may fail. The PMU placement strategy is modified to have redundancy of 1 so that the system remains observable even if one PMU is lost.

8.5.2 PMU Placement to Detect Topology Errors

Topology errors are caused by incomplete or wrong information about one or more circuit breakers at the substations. Another type of topology error is the *branch topology error*, which is due to information about the status of a given branch (i.e., whether the branch is in or out of service). Branch topology errors are easy to identify compared to the errors in several breakers. Detection of branch topology error is linked to redundancy in measurement and configuration. It is therefore possible to detect topology errors by placing meters at strategic places. One of the criterion for placing optimal PMUs can also be to improve topology error detectability.

The addition of a few PMUs for the purpose of detecting topology errors can prove to be advantageous in places where the possibility of errors in several circuit breakers exists. Although by adding few PMUs it may be possible to detect all branch topology errors, but if we cannot install enough PMUs it may be possible to detect topology errors in certain branches, which surely can only be improvement in state estimation. The method can be explained as

- Initially identify all the branches with topology errors that cannot be detected;
- For each identified branch having a topology error, determine candidate PMUs that will make the branch topology error detectable;

- Use the optimal PMU placement program to determine the minimum number of PMUs.

8.5.3 Placement of PMUs for Bad Data Detection

As discussed in Chapter 5, there are a number of methods for detecting and identifying bad data in a set of measurements. One conclusion that can be easily drawn from these observations is that if there is not enough redundancy in measurements at certain locations, the bad data cannot be detected. A system with low redundancy may contain several critical measurements. Bad data associated with critical measurements cannot be detected or identified.

Critical measurements can be identified by well-documented procedures using topological or numerical methods already described earlier in the book. Once critical measurements are identified, they can be transformed into redundant measurements by adding a few measurements at the strategic locations. Thus PMUs can be placed at strategic locations to convert the critical measurements into redundant ones. The problem of placing the PMUs is tackled in two stages. In the first stage candidate PMU locations that will transform each identified critical measurement into redundant measurement is determined. In the second stage optimal selection is made out of these candidates to find the optimal solution.

In order to determine the critical measurements and candidate measurements to make them redundant, the following method can be used:

- Form the Jacobian measurement matrix H.
- Augment this matrix by adding all the candidate PMUs at the bottom. The Jacobian measurement H can be partitioned into two submatrices as

$$H = \begin{bmatrix} H_{used} \\ H_{PMU} \end{bmatrix}$$

  where H_{used} rows correspond to existing measurements and H_{PMU} rows correspond to PMU measurements. Here the system is assumed to be already observable.
- Perform a rectangular $L\,U$ factorization on H using the Peters-Wilkinson method. Now critical measurements are identified as described earlier.

- Once the critical measurements are identified, a set of candidate PMUs is selected for each critical measurement. These candidates are chosen such that any one of them will transform the corresponding critical measurement into a redundant one.
- In the final step for optimal selection, a minimum required number of PMUs from the list of candidates identified are determined using the optimal placement programs discussed in Section 8.5.
- For mixed measurements where the candidate measurements set will also include the conventional measurements, the Jacobian matrix will be modified as

$$H_{cand} = \begin{bmatrix} H_{PMU} \\ H_{conv} \end{bmatrix}$$

and the Jacobian measurement will be

$$H = \begin{bmatrix} H_{used} \\ H_{PMU} \\ H_{conv} \end{bmatrix}$$

The same procedure will be used to determine the critical measurements and from the candidate PMUs an optimal selection will be made.

In this section, it is shown that by adding few PMUs in the system that is initially observable, it is possible to detect and identify bad data in the low redundancy portion of a power system.

8.6 Some Observations

In this chapter, application of a PMU in a power system state estimation has been discussed. One observation that can generally be made about PMUs is that they provide more accurate measurements than conventional measurements. It is therefore a preferable technique instead of traditional state estimators. Presently it is not possible to provide PMU measurement in every bus due to the cost and therefore hybrid state estimators are used. Although PMUs are more accurate, they are also prone to errors due to the presence of instrument transformers at the input. The errors introduced by the instrument transformers can be eliminated by calibrating them using several scans of measurements during daily load variations, provided there are a few accurate transformers available in the power system. In most power systems, accurate instrument transformers for metering purposes are always available and can be used for this purpose. A simple program for optimal placement of PMUs is also described.

The main function of PMUs in a real-time control system is in state estimation, but these can also help in topology error and bad data detection and correction. Although not discussed here, PMUs are also useful in other applications such as system protection, control, and stability assessment.

References

[1] Phadke, A. G., J. S. Thorpe, and K. J. Karimi, "State Estimation with Phasor Measurements," *IEEE Transactions on Power Systems,* Vol. 1, 1986, pp. 233–238.

[2] Phadke, A. G., and J. S. Thorpe, *Synchronized Phasor Measurements and Their Applications,* Boston, MA: Springer 2008.

[3] Zhou, M., V. A. Centeno, and J. S. Thorp, et al., "An Alternative for Including Phasor Measurements in State Estimators," *IEEE Transactions on Power Systems,* Vol. 2, No. 4, 2006, pp. 1930–1937.

[4] Zhao, L., and A. Abur, "Multi-Area State Estimation Using Synchronized Phasor Measurements," *IEEE Transactions on Power Systems,* Vol. 20, No. 2, 2005, pp. 611–617.

[5] Jiang, W., V. Vittal, and G. T. Heydt, "A Distributed State Estimator Utilizing Synchronized Phasor Measurements," *IEEE Transactions on Power Systems,* Vol. 22, No. 2, 2007, pp. 563–571.

[6] Kekatos, V., G. B. Giannakis, and B. Woolenberg, "Optimal Placement of Phasor Measurement Units via Convex Relaxation," *IEEE Transactions on Power Systems,* Vol. 27, No. 2, 2012, pp. 1–10.

[7] Bi, T. S., X. H. Qin, and Q. X. Yang, "A Novel Hybrid State Estimator Including Synchronized Phasor Measurements," *Electric Power Systems Research,* No. 38, 2008 pp. 1343–1352.

[8] Chen, J., and A. Abur, "Placemnet of PMUs to Enable Bad Data Detection in State Estimation," *IEEE Transactions on Power Systems,* Vol. 21, No. 4, 2006, pp. 1608–1615.

[9] Zivanovic, R., and C. Cairns, "Implementation of PMU Technology in State Estimation: An Overview," *Proc. IEEE AFRICON,* Vol. 2, 1996, pp. 1006–1011.

[10] Naqui, R. F., and A. G. Phadke, "Phasor Measurement Placement Techniques for Complete and Incomplete Observability," *IEEE Transactions on Power Delivery,* Vol. 20, No. 4, 2005, pp. 2381–2388.

[11] Gou, B., "Generalized Integer Linear Programming Formulation for Optimal PMU Placement," *IEEE Transactions on Power Systems,* Vol. 23, No. 3, 2008, pp. 1099–1104.

[12] IEEE Standard for Synchrophasors for Power System, IEEE Standard 1344-1995 (R2001), 1995.

[13] Baldwin, T. L., L. Mili, and M. B. Boisen, et al., "Power System Observability with Minimal Phasor Measurement Placement," *IEEE Transactions on Power Systems,* Vol. 8, No. 3, 1993, pp. 1099–1104.

[14] Emami, R., and A. Abur, "Robust Measurement Design by Placing Synchronized Phasor Measurements on Network Branches," *IEEE Transactions on Power Systems,* Vol. 25, No. 1, 2010, pp. 38–43.

[15] Chen, J., and A. Abur, "Enhanced Topology Error Processing via Optimal Measurement Design," *IEEE Transactions on Power Systems,* Vol. 23, No. 3, 2008, pp. 845–885.

9

Distribution System State Estimation

9.1 Introduction

The electric power system consists of generation, transmission, and distribution systems. The distribution system is the low voltage part of the system connecting consumers to the supply substation. The advances in distribution control technology are not as developed as they are for the transmission network. During the last few years the addition of distributed generation requires monitoring and control of the distribution network through a distribution management system (DMS). Two main parts of DMS are state estimation and control scheduling. State estimation techniques for transmission network analysis were described in earlier chapters of this book. The application of state estimation techniques in a distribution system is relatively new and practical applications are gradually coming. Now due to the liberalization of the energy market and increased environmental awareness, there is a tendency of having increased distributed generation connected to the distribution system. In earlier systems where centralized generation was common, the power flow was in one direction—from the generating station to the consumers. However, when there are many small generators connected to the distribution grid, the direction in which the power flows cannot be determined easily. This is the case during normal working conditions as well as during fault conditions, which may also involve safety issues. The use of state estimation in a distribution system can provide a solution to this problem.

The distribution system is lower voltage part of power system (less than 33 kV) that delivers power to the consumers from a distribution substation.

The state of a distribution grid, like the state of a transmission system, is defined by the voltage magnitude and phase angle at every node of the grid. If these values are known then the value and direction of power flow in every branch can be determined. Since in the state estimation a large number of measurements is required that are not so common in a distribution system, estimation techniques are being developed with a minimum number of data. Another feature that differentiates the estimator for a distribution system from that of a transmission system is that unlike the transmission system, the three phases of the distribution system are not balanced. Another difference with the transmission system is that most of the distribution systems have a radial topology [1–3].

Conventional state estimation methods, based on the WLS technique, are not suitable for many distribution systems due to ill-conditioned Jacobian and gain matrices. The distribution systems have high r/x ratio in lines, and also a very large number of nodes, for which the conventional estimators may not work satisfactorily. All these features make the distribution system state estimation different from transmission system state estimation, as described in next few sections.

9.1.1 Topology Estimation

All state estimation methods are based on the assumption that anaccurate system model is available for the network. However, in distribution system the state of some switching devices may not be known or the current status of all the switches in the database may not be accurate. Such conditions result in a topological error in the network. Topology processing algorithms have been developed to detect network topology errors efficiently and accurately. The main goal of a network topology processor (NTP) is to generate and maintain the power system model (of a node-branch type), based on the real-time data collected by the SCADA system. The NTP determines both the connectivity of the electrical network and locations of metering devices in the system. Currently, available topology processors perform the following functions [4]:

- Determine node-branch network connectivity;
- Connectivity status of equipment;
- Information concerning equipment out of service;
- Open-ended lines and transformers;
- De-energized equipment;
- Network islands.

After the topology processing is complete, the state estimation is performed to determine system states and existence of any analog bad data.

A conventional network topology processor is based on a search algorithm for graphs. First, the analog measurements are converted to appropriate units and verified for operating limits and for nonzero flows in open switching devices and so forth. Next, bus sections at each substation voltage level are processed to determine if they are connected by closed switching devices. This now becomes a single bus. Other buses are associated with a terminal of branch devices such as transmission lines, transformers, shunt devices such as capacitors, and metering devices. This step is known as substation configuration. The connectivity among substations is analyzed to construct a model for the entire network known as network configuration.

Normally, a basic width-first searching or depth-first searching algorithm in graph theory is applied to identify bus and network configurations. This is achieved in two steps. In the first step, graphs with nodes as vertices and breakers and disconnectors as edges are searched, and in next step, graphs with buses and impedance branches as vertices are searched.

9.2 Pseudomeasurements

Unlike the state estimation in transmission network where a large number of measured data is available, the redundancy for distribution system is generally low [1]. To make up for that, additional nontelemetered information on system quantities known as pseudomeasurement is employed so that state estimation programs can work efficiently. One possible method of adding pseudomeasurements is to use the historical data of feeders and distribution transformers loading. Pseudomeasurements using historical data may be quite different from the real-time actual values, but they increase the redundancy of data for state estimation. It is therefore important to generate more accurate pseudomeasurements by using accurate modeling of loads. A number of methods are now available for modeling of loads to improve the accuracy of pseudomeasurements for state estimation. Some of these methods are described briefly [5–7].

From the historical load data, the probability density function (PDF) of a load for a specific time period can be plotted, as shown in Figure 9.1. The randomness of the customer load behavior makes it necessary to use statistical methods of variance, normal distribution fitting, and Gaussian mixture modeling for the pseudomeasurement of load. The most common technique to model the loads is through Gaussian distribution. However, a single Gaussian assumption cannot be justified for all loads. During the research in modeling of loads at various places, it was found that the statistical distribution of electric load variation does not follow any common probability distribution function. Still, Gaussian modeling of a load profile is used because of its simplicity. In each case the PDF is characterized by one or more values of mean μ and variance σ^2.

9.2.1 Method of Assumed Variance (MAV)

From the histogram of the load profile shown in Figure 9.1, the measurement value μ is obtained by method of assumed variance as

$$\mu = \frac{1}{n}\sum_{i=1}^{n}\alpha_i \tag{9.1}$$

where α_i is the ith sample of the random variable demand α, and n is the size of the samples taken from the historical data. The standard deviation σ of the pseudomeasurements is assumed arbitrarily as a percentage (20% to 50%) of measurement value μ.

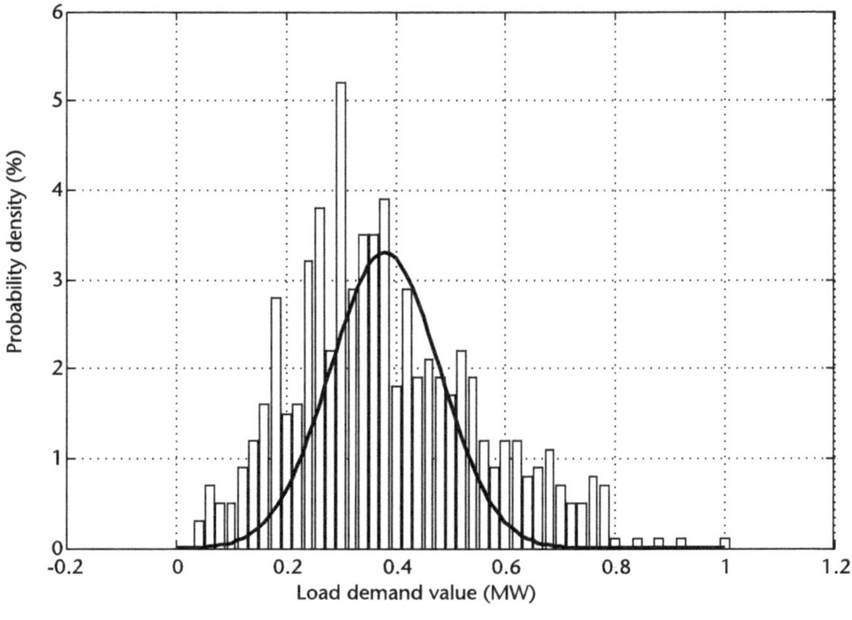

Figure 9.1 Original load PDF and the normal distribution model.

$$\sigma = \delta\mu$$

where the value of δ is 0.2–0.5.

The same value of α is assumed for all load pseudomeasurements irrespective of time and location of load. The method used in state estimation gives good results as far as voltage magnitude estimation is concerned, but the accuracy of voltage angle estimates is not so good.

9.2.2 Normal Distribution Fitting (NDF)

In this method the measurement value is also obtained using (9.1). Here the variance is calculated using the normally distributed model for the load shown in Figure 9.1.

$$\sigma^2 = \frac{1}{n-1}\sum_{-=1}^{n}(\alpha_i - \mu) \tag{9.2}$$

As can be seen from Figure 9.1, the normal distribution model does not represent the PDF of the load. The results obtained from this method are similar to the results obtained by the method of assumed variance.

9.2.3 Gaussian Mixture Model (GMM)

GMM approximation has the advantage that different types of load distribution can be fairly represented as convex combinations of several normal distributions with respective means and variances, as shown in Figure 9.2 [10]. A Gaussian mixture (GM) PDF is a weighted finite sum of Gaussian PDFs, and is characterized by the number of mixture components and the weights, mean, and variance of each component. The mixture weights have to be positive and sum of all weights must be equal to 1, as an integral of a PDF over a sample space of the random quantity must be equal to 1, and all PDFs must be positive.

The multicomponent GM PDF model is given by

$$f(\alpha) = \sum_{i=1}^{N_c} w_i f(\alpha \mid \mu_i \textstyle\sum_i) \tag{9.3}$$

Subject to $w_i > 0$, and $\sum_{i=1}^{N_c} w_i = 1$

where N_c is the number of mixture components, w_i is the weight of the mixture component, $f(\alpha|\mu_i \sum)$ is the *i*th density function of each mixture component and has a normal distribution given by

$$f(\alpha|\mu_i \textstyle\sum_i) = \frac{1}{(2\pi)^{\frac{1}{2}} \det\left(\sum_i\right)^{\frac{1}{2}}} e^{-\frac{1}{2}(\alpha-\mu_i)^T \sum_i^{-1} (\alpha - \mu_i)} \tag{9.4}$$

where μ_i is the mean of the *i*th normal mixture component, and covariance

$$\sum_i = E\left[(\alpha - \mu_i)^T (\alpha - \mu_i)\right] \tag{9.5}$$

the pseudomeasurement value μ is equal to the mean of the normal mixture component closest to the estimated demand, and σ^2 is set equal to the variance of the normal mixture component.

The expectation maximization (EM) algorithm is an efficient tool to obtain w_i, μ_i, and σ_i^2, where $i = 1, 2....$ The EM algorithm recursively generates a sequence of current parameter estimates that under normal circumstances converge to maximize the log likelihood expectation. For details of the EM algorithm, the reader may consult [3] or any book of statistics.

9.2.4 Correlation Method

The three methods—MAV, NDF, and GMM—do not depend on real-time measurements, the accuracy of these models is therefore prone to uncertainties due to weather and temperature. There may be differences between the historical data and real-time loads that cannot be accounted for in these models. In the correlation method, the historical relationship between electrical quantities and real-time measurements reflects the flexibility of the model to account for real-time network conditions.

The method of correlation seeks the linear relationship between a power system quantity X measured in real time and a load Y by analyzing the historical data as

$$\mathrm{cov}(X,Y) = E(X - \mu_x)(Y - \mu_y) \tag{9.6}$$

where μ_x and μ_y are the means of X and Y, and the covariance coefficient that has a value between −1 and 1 is given by

$$\rho_{x,y} = \frac{\text{cov}(X,Y)}{\sigma_x \sigma_y} \tag{9.7}$$

where σ_x and σ_y are standard deviation of X and Y, respectively.

Regression analysis is applied to find out the degree of dependency between the nonmonitored load and the real measurement. If the value of correlation coefficient is near to extreme values, the relationship between X and Y is linear. The correlation and GMM methods are found to give better results compared to the NAV and NDF methods.

9.3 State Estimation for Radial Distribution Systems

As discussed in Section 9.1 traditional state estimation methods suffer from convergence and/or numerical difficulties when applied to distribution systems. For a radial distribution system, a different approach has been suggested. Since the distribution systems have a large number of current measurements, their inclusion in state estimation algorithms presents problems that are well known. One of the major problems is the possibility of obtaining multiple solutions. In order to circumvent this problem, inequality constraints were introduced in state estimation solutions. The problem of state estimation in distribution systems has received the attention of researchers during the last decade. Three-phase state estimators were developed to deal with three-phase voltage magnitude measurements, per-phase power measurements, and current measurements.

The three-phase state estimation is based on the same model as used for a conventional state estimator.

$$z = h(x) + e \tag{9.8}$$

In a three-phase system the system state is selected for all three phases, measurements are made for three phases, and uses the three-phase model of the system. The three-phase state estimation technique is an extension of a robust three-phase power flow algorithm developed by Thukaram, et al. [7]. In this method the radial property of the distribution system is exploited, and forward and backward propagation techniques to calculate branch current and node voltages are applied. This method uses the observability routine to decide whether sufficient measurement is available for state estimation. Also bad data is detected and eliminated or replaced by pseudomeasurements.

Wherever the data is not available, historical load data is used to predict the loads.

The state estimation problem is solved using a method of network flow based on a walk along the network tree in forward and backward directions. In a radial power distribution system, the power is fed at a single node called the root node or parent node. A single main and several lateral feeders are part of a distribution system that ends at nodes called child nodes. For a backward walk we start at the end nodes and move backward over each section. The forward walk is started from the root node and moves forward over each section. Figure 9.3 shows the three-phase branch model for a branch.

9.3.1 State Estimation Scheme

First, the state values are initialized either to the measured values or if measured values are not available, the voltage magnitudes are set to 1.0 p.u. and phase angles for phase a, b, and c, as 0.0, 120, and −120 degrees. All branch currents and complex power are also set to 0.0. The source node is taken as a reference having a voltage of 1.0 pu and an angle of 0.0 degree.

Backward propagation. In backward propagation the branch currents and power flows are calculated in each section. The voltage values are held constant, and information about branch currents and weighted averaged

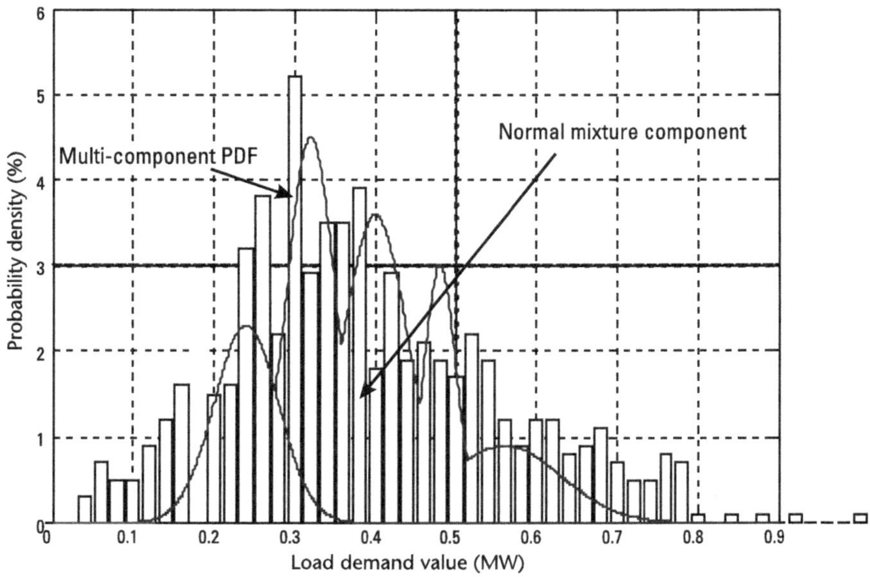

Figure 9.2 Gaussian mixture PDF.

power flow are transmitted backward along the feeder using a backward walk. During backward propagation, the load current is calculated assuming the load as demand measurement in each node, depending on the type of load.

For constant power loads, the load current at the *i*th end node is given by

$$\begin{bmatrix} I_a(i) \\ I_b(i) \\ I_c(i) \end{bmatrix} = \begin{bmatrix} \dfrac{S_a(i)}{V_a(i)} \\ \dfrac{S_b(i)}{V_b(i)} \\ \dfrac{S_c(i)}{V_c(i)} \end{bmatrix} \tag{9.9}$$

For a constant current load the load current is the same as load

$$\begin{bmatrix} I_a(i) \\ I_b(i) \\ I_c(i) \end{bmatrix} = \begin{bmatrix} I_{La}(i) \\ I_{Lb}(i) \\ I_{Lc}(i) \end{bmatrix} \tag{9.10}$$

For a constant impedance load, the load current is

$$\begin{bmatrix} I_a(i) \\ I_b(i) \\ I_c(i) \end{bmatrix} = \begin{bmatrix} \dfrac{V_a(i)}{Z_a(i)} \\ \dfrac{V_b(i)}{Z_b(i)} \\ \dfrac{V_c(i)}{Z_c(i)} \end{bmatrix} \tag{9.11}$$

where $S_a(i)$, $S_b(i)$, $S_c(i)$ are the complex powers of the constant power load, and Z_a, Z_b, and Z_c, are the impedance of the constant impedance load at bus i, respectively.

When the currents are known, the branch flow in the corresponding branch can be obtained. The current in branch m having parent node j and child node i is

$$\begin{bmatrix} I_a(m) \\ I_b(m) \\ I_c(m) \end{bmatrix} = \begin{bmatrix} I_a(i) \\ I_b(i) \\ I_c(i) \end{bmatrix} + \sum_{p\mathring{a}m} \begin{bmatrix} i_a(p) \\ i_b(p) \\ i_c(p) \end{bmatrix} \tag{9.12}$$

where branch m is connected to branch P_i and branch p has parent node I_p that is the child node of branch m.

The power flow in branch m is calculated as $P_{ij}(m) + Q_{ij(m)}$ for each phase a, b, and c. For phase a it is

$$P_{ija}(m) = Re V_a(j) I_a^*(m) \tag{9.13}$$

$$Q_{ija}(m) = Im V_a(j) I_a^*(m) \tag{9.14}$$

For other phases the power can be obtained in the same way. These power flows are weighted averaged with measured values, then the parent branch current and branch power flow is updated.

Forward propagation. In forward propagation, voltage and load at each node is calculated starting from the source node of the feeder to the end of the feeder. The branch currents are calculated based on average flow, and these values are used to calculate the node voltages and the loads at each node. The node voltage can be calculated from branch currents and branch impedances. As

$$\begin{bmatrix} V_a(i) \\ V_b(i) \\ V_c(i) \end{bmatrix} = \begin{bmatrix} V_a(j) \\ V_b(j) \\ V_c(j) \end{bmatrix} - \begin{bmatrix} Z_{aa}(m) & Z_{ab}(m) & Z_{ac}(m) \\ Z_{ba}(m) & Z_{bb}(m) & Z_{bc}(m) \\ Z_{ca}(m) & Z_{cb}(m) & Z_{cc}(m) \end{bmatrix} \begin{bmatrix} I_a(m) \\ I_b(m) \\ I_c(m) \end{bmatrix} \tag{9.15}$$

The values of branch currents are the updated values based on average weighted power flows. The steps of backward and forward propagation are followed during each iteration of voltage computations. When the values of complex voltages at all the nodes compared to the previous iteration are within a specified limit, the process is stopped.

9.4 Branch Current Based Estimation Method

A branch current based state estimation algorithm uses magnitudes and phase angles of branch currents as state variables. The benefit of using current magnitudes as state variables is that current magnitude measurements are the predominantly measurement available in distribution systems, which correspond directly with the state variable. The algorithm presented here uses the WLS method commonly used in state estimation [8].

Equation (9.8) is the general equation that will be used. The solution to this equation has already been discussed in Chapter 3. The Jacobian matrix will be modified here to include current magnitude measurements and all other quantities on a per-phase basis. The measurement equations and the Jacobian entries for different types of measurements are given below.

9.4.1 Branch Power Measurements

The power for phase a (a-b-c) from bus *j* to bus *k* at bus *k* can be written as

$$P_{jk}^{a} + Q_{jk}^{a} = V_j^a (I_{jk}^{a})^* \tag{9.16}$$

$$V_j^a I_{jk}^a \left[\cos\left(\delta_j^\alpha - \alpha_{jk}^\alpha\right) + j\sin\left(\delta_j^\alpha - \alpha_{jk}^\alpha\right)\right. \tag{9.17}$$

where P_{jk}^a, Q_{jk}^a are branch real power and reactive power of phase a from bus *j* to bus k, respectively. $I_{jk} = \left|I_{jk}\right| < \alpha$ is a complex branch current at bus *j*, and $V_j^a = \left|V_j^a < \delta\right|$ is the complex voltage of phase a at bus *j*.

When the measurements and the state variables belong to same line segment, the Jacobian matrix entries are

$$\frac{\partial P_{jk}^a}{\partial I_{jk}^a} = V_j^a \left[\cos\left(\delta_j^a - \alpha_{jk}^a\right) = \frac{P_{jk}^a}{I_{jk}^a}\right. \tag{9.18}$$

$$\frac{\partial P_{jk}^a}{\partial \alpha_{jk}^a} = V_j^a I_{jk}^a \left[\sin\left(\delta_j^a - \alpha_{jk}^a\right) = Q_{jk}^a\right. \tag{9.19}$$

$$\frac{\partial Q_{jk}^a}{\partial I_{jk}^a} = V_j^a \left[\sin\left(\delta_j^a - \alpha_{jk}^a\right) = \frac{Q_{jk}^a}{I_{jk}^a}\right. \tag{9.20}$$

$$\frac{\partial Q_{jk}^a}{\partial \alpha_{jk}^a} = -V_j^a I_{jk}^a \left[\cos\left(\delta_j^a - \alpha_{jk}^a\right) = -P_{jk}^a \qquad (9.21)$$

When the measurements and the state variables are not in the same line segment, all entries related to branch power measurements are zero.

9.4.2 Current Magnitude Measurements

For current flowing from bus *j* to bus *k* in phase *a*, the current measured is I_{jki}^a and the Jacobian matrix entries are $\frac{\partial I_{jk}^a}{\partial I_{jk}^a} = 1$. For other measurement that are not in the same line as the state variables the entries are 0. Also

$$\frac{\partial I_{jk}^a}{\partial \alpha_{jk}^a} = 0 \qquad (9.22)$$

9.4.3 Power Injection Measurements

There are two conditions in power injection at any bus. Suppose there are n buses connected to bus j, the direction of current flow towards bus j is from m buses , and in the remaining $m + 1$ to n buses the current flows from bus j to these buses. The power injected at bus j in phase a is

$$P_j^a + jQ_j^a = V_j^a \left[\sum_{i=1}^{m} I_{ij}^a - \sum_{i=m+!}^{n} I_{ji}^a \right] \qquad (9.23)$$

The corresponding Jacobian matrix entries are

1. When line segment is connected to a bus at which power is injected and the state variables are for the buses 1 ... m. For state variables and the measurements in the same phase, the entries are

$$\frac{\partial P_j^a}{\partial I_{ij}^a} = V_j^a \cos\left(\delta_j^a - \alpha_{ij}^a\right)$$
$$\frac{\partial P_j^a}{\partial \alpha_{ij}^a} = V_j^a I_{ij}^a \sin\left(\delta_j^a - \alpha_{ij}^a\right)$$
$$\frac{\partial Q_j^a}{\partial I_{ij}^a} = V_j^a \sin\left(\delta_j^a - \alpha_{ij}^a\right) \tag{9.24}$$
$$\frac{\partial Q_j^a}{\partial \alpha_{ij}^a} = -V_j^a I_{ij}^a \cos\left(\delta_j^a - \alpha_{ij}^a\right)$$

If measurements and state variables are not in same phase, entries in the Jacobian matrix are zero.

2. When a line segment is connected to a bus at which power is injected and the state variable are for the buses $m + 1$ to n. If the measurements and state variables are in the same phase, the Jacobian matrix entries are

$$\frac{\partial P_j^a}{\partial I_{ij}^a} = -V_j^a \cos\left(\delta_j^a - \alpha_{ij}^a\right)$$
$$\frac{\partial P_j^a}{\partial \alpha_{ij}^a} = -V_j^a I_{ij}^a \sin\left(\delta_j^a - \alpha_{ij}^a\right)$$
$$\frac{\partial Q_j^a}{\partial I_{ij}^a} = -V_j^a \sin\left(\delta_j^a - \alpha_{ij}^a\right) \tag{9.25}$$
$$\frac{\partial Q_j^a}{\partial \alpha_{ij}^a} = V_j^a I_{ij}^a \cos\left(\delta_j^a - \alpha_{ij}^a\right)$$

If measurements and state variables are not in the same phase, the entries are zero.

9.4.4 Voltage Magnitude Measurements

In the case of voltage magnitude measurements, there may be a line segment that will be falling between the bus at which voltage is measured and the source bus and there will be entries between measurements and state variables of different phases. However, it is found that these values are very small and can be approximated to zero. That is, the three phases are assumed to be decoupled.

The algorithm can be implanted in the following manner:

1. Initialization is performed using backward propagation and calculating the branch current from bus node voltage initialized to 1.0 p.u. Next, forward propagation is used to calculate the node voltages.
2. In the next step, the states of the system (branch currents) for the three phases are calculated.
3. State variables are updated.
4. Check for the difference from the previous value. If within the limits specified, then stop.

9.5 Object-Oriented State Estimation

Object-oriented programming (OOP) technology is a new methodology that is being applied extensively in areas of software development. Using the OOP approach, the programs can be written in a more general way instead of using application-oriented programs. The modern distribution management systems (DMSs) are making use of OOP, since it offers more flexibility, expansibility, and easy maintenance and upgrade. Initially, the OOP in DMS was developed for load flow for radial distribution systems. Everything in OOP is grouped as self-sustainable *objects*. An object here is considered as a thing that can perform a set of related activities. In pure OOP terms an object is an instance of a *class*. The key concept of object-oriented modeling of a system is the class, a programming entity that represents the set of objects with similar properties and behavior. In a radial power distribution system the power is fed at a single node called the root node. A single main and several lateral feeders are part of a distribution system. Feeders are comprised of branches that may be either a transformer or a transmission line section. Usually a bus is connected with two branches. A fork node is a bus to which more than two branches are connected. A bus that is connected to only one branch is called a terminal node.

In object-oriented modeling of a distribution system, we will be dealing with physically existing components such as feeders, buses, branches, source, load, and shunt devices that will be modeled as software objects. The distribution system itself can be modeled as an abstract object.

For state estimation the base class has ports, port variables, and a computational method. A port here will be defined as ingoing or outgoing depending on the direction of power at the port. For balance of a three-phase system in steady state the port will be characterized by four variables: real and imaginary components of voltage, and real and reactive power. The model of a distribution system is an oriented graph of objects appropriately connected.

The WLS method of state estimation with constrained nonlinear minimization problem can be written as

$$\min J(x) = \left[z - h(x)\right]^T \left[R^{-1}\right]\left[z - h(x)\right] \tag{9.26}$$

subject to $c(x) = 0$
where $c(x)$ is the r-vector of the equality constraint nonlinear functions. Other symbols have usual meanings, $r \le n$, $n \le m = r$. The backward/forward sweep method described earlier can be easily implemented in an OO framework.

9.6 Measurement Placement

During the last few years, a considerable portion of generation is connected at the distribution level (distributed generation). This requires the monitoring and control of the power system by means of a modern DMS at the substation level. The state estimation for DMS has to deal with a large portion of unmonitored sections particularly at low voltage level. Voltage, current, and power flows are measured at the primary distribution voltage level in primary substations. The secondary level distribution are virtually not monitored. Loads are not measured but modeled as pseudomeasurements. Thus the estimated values of voltage magnitude and phase angle are not very accurate. In order to improve the accuracy of estimates, it is therefore necessary to employ more direct measurements. Since the cost of providing measurements at every location is a very costly affair, choices have to be made regarding the locations, types, and number of measurements. The placement problem has been dealt with extensively in the literature for transmission systems. At the distribution level not much literature is available. Basically, the measurement placement problem is dealt with mainly for two purposes [10–12]:

1. To improve the network observability;
2. To minimize the errors in estimation.

In order to improve the observability of the distribution network, the measurement placement techniques are similar to the ones used in a transmission system. There are three principal types of algorithms for observability analysis used in a power system: topological, numerical, and hybrid. All these techniques have been described in Chapter 4. The topological analysis or graph theoretic procedure expands a subgraph from the part of the system so far made observable into the remaining unobservable parts. If PMUs are used to improve the observability the subgraph is built according to following rules:

- Place the first PMU at the bus that has the highest number of incident lines in the unobservable region;
- Determine the observable region by this PMU;
- Check the whole system for observability. If it is not observable, add another PMU until the total system becomes observable.

The numerical observability algorithm handles networks that are totally or partially represented at the physical level. It can be categorized in two main categories:

1. Gaussian elimination or triangular factorization of the Jacobian matrix;
2. Null space of the Jacobian or gain matrix.

Not much research work is reported on improving the accuracy of states in estimation using measurement placement [9–11]. One of the methods suggested is implanting using a rule-based technique of meter placement. The first rule is to place the meters in the main switch and fuse locations for monitoring and control of the switch and feeder. Additional measurements are then installed in such manner as to have almost equal load in each meter zone. A few meters can also be provided on normal open points for the purpose of feeder switching and reconfiguration. Reconfiguration of feeders is employed under normal conditions to reduce the system losses and/or balance the loads in the feeders. During reconfiguration the switch position is changed, keeping the load flows and constraints maintained.

The set of measurements is reduced based on the accuracy index that is based on the sum of the variances of the measurements. A heuristic technique has also been applied that identifies locations to place a certain number of voltage measurements to minimize the standard deviation in voltage estimates of a nonmetered bus. In this method a predetermined number of measurements is chosen arbitrarily. These methods are not very accurate, especially in the estimation of voltage phase angles.

A probabilistic approach based on Monte Carlo simulations for optimal measurement placement has been suggested recently. Mathematically the problem of meter placement to identify the effective locations and number of real measurements can be expressed in terms of probability indices as

$$p_i = P_r \left\{ \left| \frac{\hat{V}_i \quad -V_i}{V_i} \right| \right\} \leq \epsilon_1, \left| \frac{\hat{\delta}_i - \quad \delta_i}{\delta_i} \right| \leq \epsilon_2 \tag{9.27}$$

$$i = 1, 2, 3, \ldots, n$$

where V_i, δ_i are measured values of voltage and angle at bus, $\hat{V}_i, \hat{d}_i$, δ are estimated values of voltage and angle at bus *i*, *e* is the permissible threshold error.

One of the buses is chosen as the main substation bus and taken as the reference with voltage =1.0 p.u. and angle $\delta = 0$. The objective is to bring down the relative errors in the voltage and angle estimates, and all buses, below some predefined threshold values (1% for voltage error and 5% for angle error is common).

The measurement placement problem is considered from the point of view of identifying the meter locations to bring down the errors. The obvious choice to improve the index in (9.27) 0 is to place the meters at the locations where the determinant of error covariance matrix is largest. However, locating voltage measurements only may not result in bringing the angle estimates to below threshold value. After running a few simulations, if it is found that even by placing a large number of voltage measurements the angle error is not reduced, then other measurement such as line power flow measurements may be added. The algorithm for location and number of measurement placements can be described as

1. Run the WLS state estimation program over a set of Monte Carlo simulations and determine the errors in voltage magnitudes and angles in all buses in each simulation.
2. If more than 95% of the states have errors within the threshold value, stop the simulation. If voltage magnitude errors are within the threshold value, but angles have errors more than limits, then power flow measurement placement may be required.
3. If the errors in both voltage magnitude and angle are not within the limit, voltage measurement placement may be tried.
4. For each bus draw a graph (ellipse) between the error in the voltage versus the error in the angle for all Monte Carlo simulations. At

every bus, compute the area of error ellipse, identify the bus with the largest area, and place the voltage measurement at this bus. If a measurement is already present select the bus with the next largest area.

5. Go back to step 1 until the errors are within the limit.
6. If angle errors are not within the limit but the voltage errors are, then draw the graph between the line flow error and angle error, which is again an ellipse. Place the flow measurement in the line with the largest area. If the measurement is already present, choose the next line having the largest area.
7. Go back to step 1.

The flow chart of the measurement placement technique is shown in Figure 9.3.

9.7 State Estimation with Kalman Filter

The Kalman filter is a state estimator that produces an optimal state of a process such that it minimizes the mean of the squared error. The filter is very powerful in several aspects: it supports estimation of past, present, and even future states, and it can do so even when the precise model of the system is not known. The Kalman filter has an iterative structure and two types of states. There is a corrector step where the most recent measurement is taken into account, and there is a prediction step for the next time step [13].

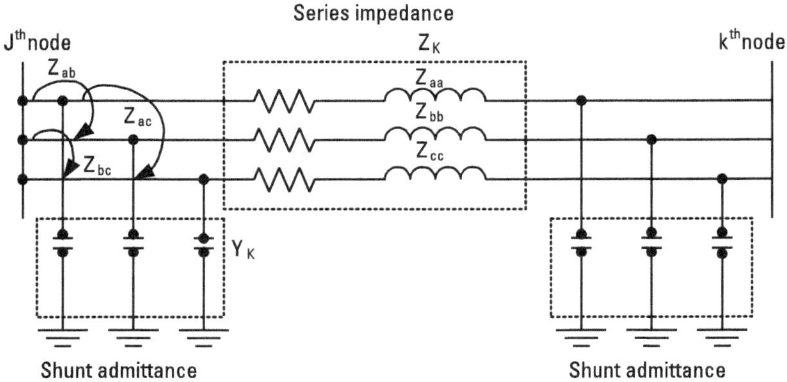

Figure 9.3 Three-phase model of line section.

The Kalman filter assumes that the system model consists of a discrete time nonlinear state space model written as

$$x(k+1) = f\{x(k), u(k) + G w(k) \tag{9.28}$$

and the nonlinear measurement model

$$z(k) = h\{x(k), u(k) + H w(k) + v(k) \tag{9.29}$$

where $[x]^T = [x_1, x_2, \ldots, x_n]$ is the state vector, $[u]^T = [u_1, u_2, \ldots, u_m]$ is the input vector, z is the measurement vector of r measurement variables, $f(\ )$ is the system vector function, w is the noise vector, G is the noise gain matrix relating noise to state variables, $h(\ )$ is the measurement vector function, H is the gain matrix relating disturbances to measurements directly, and v is the random measurement noise vector.

The Kalman filter state estimation algorithm for distribution system is as follows:

- *Initializaion*: The initial value of the predicted state estimate $x_p(0)$ is initialized to $V = 1.0$ p.u. and angle $\theta = 0$.
- *Prediction step:* Calculate the predicted measurement state estimate from the predicted state estimate.

$$z_p(k) = hx_p(k) \tag{9.30}$$

- Calculate the error vector as the difference between the measurement $z(k)$ and predicted measurement $z_p(k)$.

$$e(k) = z_k - z_p(k) \tag{9.31}$$

- *Corrective step:* In the corrective step the corrected value of state estimate $x_c(k)$ is obtained as

$$x_c(k) = x_p(k) + Ke(k) \tag{9.32}$$

  where K is the Kalman filter gain.
- *Prediction step:* Calculate the predicted state estimate for the next step using the present state estimate

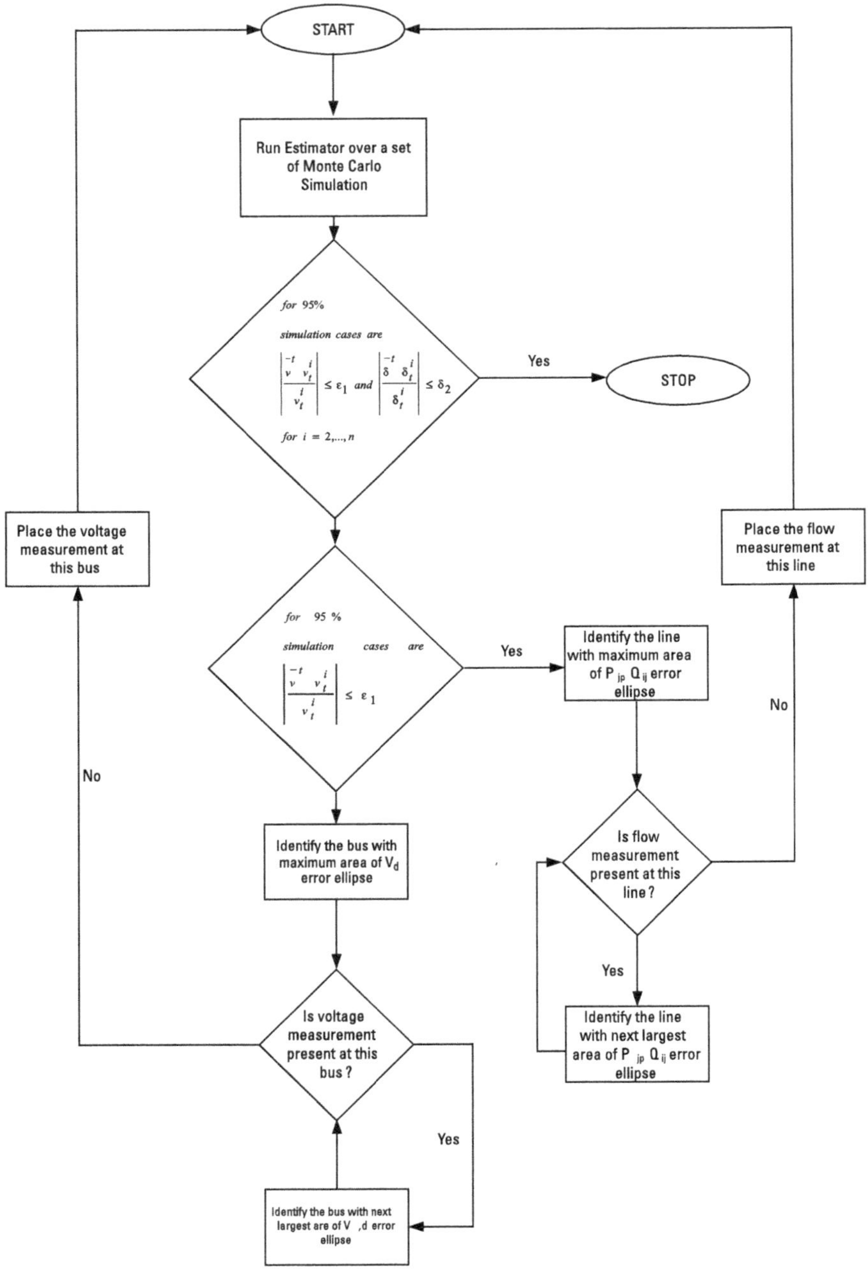

Figure 9.4 Flow chart of measurement placement techniques.

$$x_p(k+1) = f\{x_c(k), u(k)\} \tag{9.33}$$

9.7.1 Test System

The state estimation using a Kalman filter is obtained for the IEEE 34-bus system and the results are compared with the load flow solution, as shown

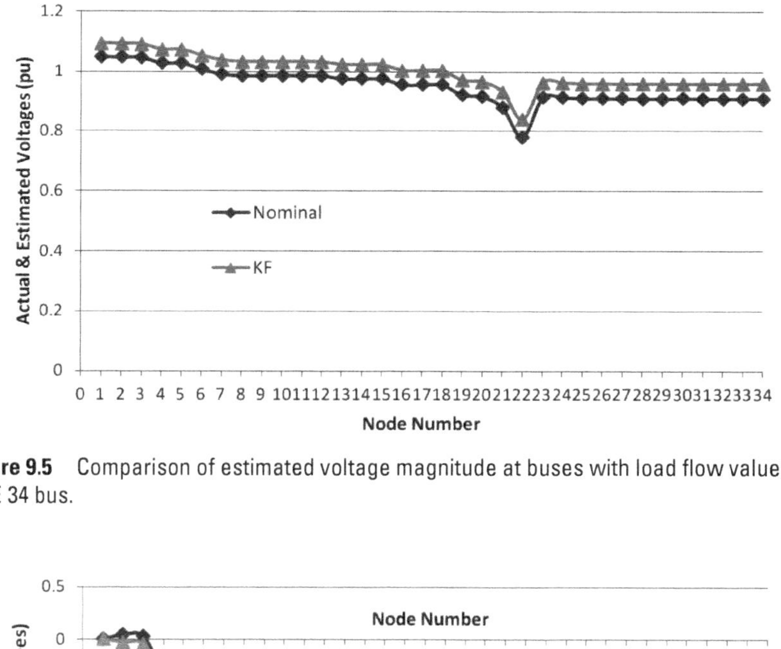

Figure 9.5 Comparison of estimated voltage magnitude at buses with load flow values for IEEE 34 bus.

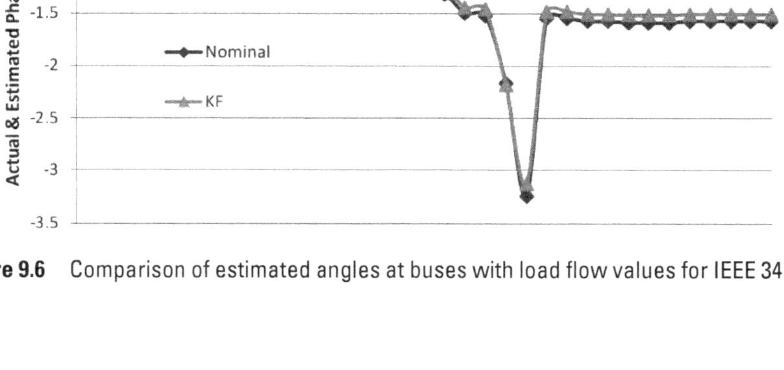

Figure 9.6 Comparison of estimated angles at buses with load flow values for IEEE 34 bus.

in Figures 9.5 and 9.6. The results obtained from the Kalman filter are very close to the actual values obtained from the load flow.

References

[1] Lu, C. N., J. H. Teng, and W. H. E. Liu, "Distribution System State Estimation," *IEEE Transactions on Power Systems,* Vol. 10, No. 1,1995, pp. 229–240.

[2] Kersting, W. H., "Distribution System Modeling and Analysis," Boca Raton, FL: CRC Press, 2002.

[3] Heydt, G. T., "The Next Generation of Power Distribution Systems," *IEEE Transactions on Smart Grid,* Vol. 1, No. 3, 2010, pp. 225–235.

[4] Prais, M., and A. Bose," A Topology Processor That Tracks Network Modifications Over Time," *IEEE Transactions on Power Systems,* Vol. 3, No. 3, 1988, pp. 992–998.

[5] Singh, R., B. C. Pal, and R. A. Jabr, "Statistical Representation of Distribution System Loads Using Gaussian Mixture Model," *IEEE Transactions on Power Systems,* Vol. 25, No. 1, 2010, pp. 29–37.

[6] Baran, M. E., N. C. Raleigh, and A. W. Kelly, "State Estimation for Real Time Monitoring of Distribution Systems," *IEEE Transactions on Power Systems,* Vol. 9, No. 3, 1994, pp. 1601–1609.

[7] Thukaram, D., J. Jerome, and C. Surapong, "A Robust Three-Phase State Estimation Algorithm for Distribution Networks," *Electric Power System Research,* Vol. 55, 2000, pp. 191–200.

[8] Wang, H., and N. N. Shultz, "A Revised Branch Current Based Distribution System State Estimation Algorithm and Meter Placement Impact," *IEEE Transactions on Power Systems,* Vol. 19, No. 1, 2004, pp. 207–213.

[9] Casolino, G. M., and A. Losi, "An Object Oriented State Estimation for Radial Distribution Systems," *Engineering Intelligent Software,* Vol. 11, No. 4, December 2003, pp. 193–200.

[10] Singh, R., B. C. Pal, and R. B. Vinter, "Measurement Placement in Distribution System State Estimation," *IEEE Transactions on Power Systems,* Vol. 24, No 2, 2009, pp. 668–675.

[11] Li, K., "State Estimation for Power Distribution System and Measurement Impacts," *IEEE Transactions on Power Systems,* Vol. 11, No. 2, 1996, pp. 911–916.

[12] Baran, M. E., N. C. Raleigh, and A. W. Kelly, "Meter Placement for Real Time Monitoring of Distribution Feeders," *IEEE Transactions on Power Systems,* Vol. 11, No 1, 1996, pp. 332–337.

[13] Blood, R., M. Ilic, and B. Krogh, et al., "A Kalman Filter Approach to Quasi-Staitc State Estimation in Electric Power Systems," *Proceedings of the 38th North American Power Symposium,* Carbondale, IL, 2006.

About the Author

Mukhtar Ahmad is a professor of electrical engineering at Aligarh Muslim University, Aligarh, India. He has 40 years of classroom teaching experience of undergraduate and graduate students at universities in India and Malaysia. He is actively engaged in research in the area of power system control, state estimation, and smart grids. He has also worked in the area of power electronics. He has published three books on electric drives and a large number of papers in international journals in these areas, and has presented papers in many countries.

Index

Recent Artech House Titles in Power Engineering

The Advanced Smart Grid: Edge Power Driving Sustainability, Andres Carvallo, and John Cooper

Battery Management Systems for Large Lithium Ion Battery Packs, Davide Andrea

Designing Control Loops for Linear and Switching Power Supplies: A Tutorial Guide, Christophe Basso

Electric Systems Operations: Evolving to the Modern Grid, Mani Vadari

Energy Harvesting for Autonomous Systems, Stephen Beeby and Neil White

Power Line Communications in Practice, Xavier Carcelle

Power System State Estimation, Mukhtar Ahmad